TROUBLESOME RISING

TROUBLESOME RISING

A THOUSAND-YEAR FLOOD IN EASTERN KENTUCKY

Edited by Melissa Helton

FIRESIDE INDUSTRIES

Published by Fireside Industries
An imprint of the University Press of Kentucky

Copyright © 2024 by The University Press of Kentucky
All rights reserved.

Editorial and Sales Offices: The University Press of Kentucky
663 South Limestone Street, Lexington, Kentucky 40508-4008
www.kentuckypress.com

Cataloging-in-Publication data is available from the Library of Congress.

ISBN 978-1-950564-42-2 (hardcover : alk. paper)
ISBN 978-1-950564-43-9 (pbk. : alk. paper)
ISBN 978-1-950564-40-8 (epub)
ISBN 978-1-950564-41-5 (pdf)

This book is printed on acid-free paper meeting
the requirements of the American National Standard
for Permanence in Paper for Printed Library Materials.

Manufactured in the United States of America.

To all the Appalachian writers who came
before us and all who follow.

To all those who faced the waters and all who will.

Contents

List of Illustrations — xiii

Introduction — 1

Prologue

Sonja Livingston
Noah's Wife — 17

I. The Map Keeps Changing

Jesse Graves
Hanktum — 23

Julia Watts
After This, the Deluge — 24

Bernard Clay
they freaks of nature — 30

Maurice Manning
Blue Hole — 32

Leatha Kendrick
Invisible, Essential — 34

Christopher McCurry
It's Raining This Week — 40

Annette Saunooke Clapsaddle
Flood Walking — 42

Contents

Patricia L. Hudson
Forty Years and a Flood — 48

Neema Avashia
Fight from Away — 53

Melva Sue Priddy
The Shape Water Takes — 60

Wendell Berry
Making It Home — 62

Maurice Manning
Planting Trees in God's Country — 64

Amelia Kirby
Collective Healing — 66

Darnell Arnoult
This Too Is Creation — 72

II. Water's Dark Body

Marianne Worthington
Rise and Fall: A Sonnet — 75

Kari Gunter-Seymour
Coal+iron+natural gas — 76

Tina Parker
How to Sleep — 78

Carter Sickels
Troublesome Rising — 79

Jesse Graves
From the Tennessee Side of the Mountains — 85

Jamey Temple
Fire and Rain — 86

Contents

Jayne Moore Waldrop
Before . . . 89

Patsy Kisner
Flood-Watching Instructions . . . 92

Meredith McCarroll
Belonging . . . 93

Lyrae Van Clief-Stefanon
Strangers: Flood Crossing . . . 98

Robert Gipe
Wall of Water: A Story . . . 100

Annie Woodford
One of the Sounds of Water . . . 102

Melva Sue Priddy
From Furman . . . 103

Tia Jensen
Turtletalk . . . 106

Mandi Fugate Sheffel
What Water Can't Erase . . . 113

Darnell Arnoult
Knowing . . . 118

G. Akers
Stag: A Story . . . 119

Pauletta Hansel
Aerial View of Catastrophic Flooding in Eastern Kentucky . . . 121

Randi Ward
Aquarius . . . 123

Melissa Helton
Chain of Custody . . . 124

Contents

Pamela Hirschler
With What Remains   127

III. These Sunken, Unpeopled Streets

Kelli Hansel Haywood
Backwash   131

Shelly Jones
Floods Make Fertile Ground   136

Monic Ductan
The Gift Horse: A Story   143

Bernard Clay
elegy for an eastern kentucky grocery   148

Randi Ward
Dark Waters   150

Marc Harshman
Headlines and History   152

Doug Van Gundy
The Flooded Town   155

Leah Hampton
Absence and Elements, a Prayer   156

Pauletta Hansel
No Friends of Coal   162

Christopher McCurry
Brothers 12 & 14   165

Frank X Walker
Elvis   167

Nikki Giovanni
Where Was the Music   169

Lee Smith
River Rising   170

Annie Woodford
Five-Hundred-Year Rains — 173

Tina Parker
Learning from Home — 175

Savannah Sipple
Ain't No Grave: A Story — 176

Sonja Livingston
Reliquiae Diluvianae (Relics of the Flood) — 186

Scott Honeycutt
The Only Prayer — 188

George Ella Lyon
Don't Tell Me — 189

IV. There Is Nothing Untouched

Lisa J. Parker
Surge — 193

Doug Van Gundy
The Morning After — 195

Ouita Michel
Around the Table — 197

Jane Hicks
Mr. Still's Hat — 202

Erin Miller Reid
What We Saved — 204

Elizabeth Lane Glass
Aching for Troublesome — 210

Kari Gunter-Seymour
Bluegrass Navy — 215

Courtney Lucas
Mud: A Story — 217

Contents

Silas House
Pulled from the Flood — 222

Wes Browne
From the Road — 228

Amanda Jo Slone
Salvage — 234

Scott Honeycutt
To Ask — 237

Amy Le Ann Richardson
It Is Still Here, the Magic — 238

Richard Hague
After — 245

Jim Minick
After the Flood, After the Tornado, and Before the Next — 251

Epilogue

Nickole Brown
Rise — 261

A Final Note — 281

Acknowledgments — 283

Appendixes

1. Map of Hindman Settlement School Campus — 289
2. 45th Appalachian Writers' Workshop Schedule — 291
3. Historical Time Line: Hindman Settlement School — 295

Contributors — 297

Illustrations

Downtown Hindman floodwaters	19
Flooded cars	28
Josh Mullins	38
View of flooded Hindman	47
Moses Owens	70
Damaged footbridge	74
Frogtown ducks	105
Restoring journal pages	111
The Hindman kitchen	112
Mullins Center damage	117
FEMA deployment planning	130
Donations arriving	141
Jethro Amburgey Bridge	151
Uncle Sol's Cabin	161
The cleanup begins	172
Archive rescue in the Gathering Place	185
Restored archive photographs	187
Donated supplies arriving	192
North 160	214

Illustrations

Donated supplies in the Great Hall	227
Crisis therapy dogs	236
Southern Appalachian Writers Cooperative Workshop	250
Community Gather & Grow	257
Campus Map of Hindman Settlement School	290
Time Line: History of Hindman Settlement School	296

Introduction

My father died of cancer when I was in my twenties. It was his second go-around, and thankfully it went quickly, so he didn't suffer much. One month after finding it again, he stayed twelve days in hospice and was gone. He didn't want a funeral, so a few months later we had a big party as a memorial.

In the interim, I led the way in making a memorial book so we could give a copy to everyone at the party. I solicited photos, stories, snippets, and memories from folks across the incarnations of his life. It was part of the grieving process, and that little book was my eulogy. Here were things people wanted to remember about him, spiral-bound and in the physical world. It was a tangible thing to hold and keep now that his body was ash.

But, of course, that book isn't him. It doesn't embody what it was like to be his daughter. It doesn't let people truly know or understand him. It offers a few important reference points on a map that can't lead to a specific place.

That's what this anthology is. It gives glimpses. It catalogs and documents. It questions and memorializes. But while it can help readers imagine what it was like to be there before, during, and after the flood, it can't put them there. It can't even put the writers back into their own bodies. That us who experienced the flood in the highly unique ways we each did floated off, drowned, washed up somewhere far downriver, molded into mush. Of course, this is true of every moment. I can't be me from yesterday, let alone July 28, 2022. And we know this. But we can't help ourselves.

We want to capture these time-stamp life moments in amber so we can own, examine, ponder, revisit, quarantine, and share them. We capture them in photos, in videos, in body-sense memories of that water's particular smell, and in abstract word squiggles in our notebooks. And that is the valiant human effort: to pin down the intangible and hold still the ephemeral.

It is the engine of art.

We see in these pieces detailed documentation of rainfall, death tolls, and the times a person moved from this porch to that porch. We see yesteryear's floods and fictional floods. We see metaphoric floods of diaspora, pollution, and COVID. We see the terrors of extractive industry and climate change disasters. We see history and memory overlapping older history and memory. We see pointed fingers, open hands, and clenched fists. They all have mud on them.

Throughout, we ride the pattern of pieces marking the setting. *It was July 28, and I was at Hindman Settlement School in the Furman building, and it was raining.* It was *x*, and I was at *y*, and *z* was happening. Marking the time and place is an act of grounding, contextualizing, and legitimizing. Like we do when we tell stories of where we were on the morning of 9/11. Or when someone asked us to marry them. Or when we saw a famous person. Or when we found out a friend had died.

And what is Appalachian storytelling without marking the time and place? The stories we tell can't exist in a nowhere setting. It has to be in *this* specific holler in winter. Under *this* bunch of hemlocks in 1975. In *this* mamaw's kitchen on Easter Sunday. With *this* passel of writers at the Hindman Settlement School during *this* unprecedented summer rainstorm.

Hindman Settlement School was founded in southeastern Kentucky by May Stone and Katherine Pettit in 1902 as the first rural settlement school in the country. While no longer providing residence or the primary education for regional students, the Settlement School is now a nonprofit that offers four main programs to the community: dyslexia and literacy tutoring, foodways and community gardening, traditional arts, and literary arts—a program that includes the Appalachian Writers' Workshop, which many pieces in this collection reference. Hindman is the county seat of Knott County.[1]

According to the National Oceanic and Atmospheric Administration (NOAA), upward of sixteen inches of rain fell over Clay, Owsley,

Breathitt, Leslie, Perry, Letcher, and Knott Counties between July 25 and 30, triggering massive flooding throughout the region. The Carr Creek Lake site in Knott County reported 13.2 inches of rain in forty-eight hours. There is less than a one-in-one-thousand chance for this much rain to fall in a single place within a given four days.[2]

NOAA records the flooding of the Red River; the Licking River; the North, Middle, and South Forks of the Kentucky River; and the main stem of the Kentucky River where all the forks join. The North Fork in Whitesburg, Letcher County, reached a 21-foot flood height before the gauge failed. This was more than six feet above the previous flood record, set in 1957. The North Fork in Jackson, Breathitt County, hit a new record height of 43.47 feet, surpassing the 1939 record of 43.1 feet. It is staggering to think of a 43-foot flood twice in one generation.

That day, Kentucky governor Andy Beshear requested a disaster declaration from the federal government, and the next day, on July 29, President Joe Biden declared the storms and resultant floods and landslides a major disaster event.[3]

Our campus, like much of the land affected by the flood, is steep on either side of the waterway (which is one reason so many people live along the waterways and not on higher, less floodable, but also less livable ground). The bed of Troublesome Creek, which eventually joins the North Fork, is 1,010 feet above sea level here. Our two buildings farthest from each other on either side of the creek, the Preece building and Mr. Still's house, are at an elevation about one hundred feet higher than Troublesome. To illustrate the steepness of our campus, those two structures are about eight hundred feet apart as the crow flies. A squirrel trying to get from one to the other would go down a hundred feet, cross Troublesome, and go back up a hundred feet in roughly the distance of two and a half football fields. A human walking or driving along a road would travel half a mile to cross those eight hundred feet.

The flash flood tore through our campus during the early morning hours of July 28. Our section of Troublesome Creek is usually six to eight inches deep, or less if we're in a dry spell. Weeks after the flood, some of the writers who were back on campus volunteering in the community helped the staff stretch a string level from the high-water mark still on the door window of Uncle Sol's Cabin, across the iconic walk bridge, and to the opposite bank. We then dropped a line from the string down to the water. The creek had risen ten feet to the top of its bank and then ten more, transforming six inches of water into a raging twenty feet. In the Mike Mullins Center, it destroyed our offices, ground-floor apartments, archives, and storage. It inundated and damaged our hoop houses, farmers' market pavilion, Uncle Sol's Cabin, historic Combs House, ground-floor classrooms in the James Still Building, and the School of Craft's weaving studio.

Besides the steep terrain, another factor that exacerbated the flood damage was the debris. Will Anderson, our executive director, heard that a box truck had swept into the creek and backed up the water even more. Once a few trailers, sheds, vehicles, and bridge culverts were thrown in, they were bound to add to the flood height.

The water had nowhere to go except up.

We were in the midst of the Forty-Fifth Appalachian Writers' Workshop (AWW), the premier gathering of writers in our region.[4] We had eighty-five in attendance that week, and over sixty were staying on campus when the flash flood tore through. Several lost vehicles and personal belongings. Staff member Corey Terry's wife, Liz, broke her leg while they and their small children fled in the dark to higher ground. Staff members Josh Mullins and Sarah Kate Morgan were in our office spaces when the doors literally burst inward and water rushed inside, quickly rising to their chests. Everyone survived. And everyone carried some level of stress or trauma away from the experience. The remaining days of AWW were of course canceled, including a planned memorial for Ron Houchin, a longtime workshop attendee who had recently died, and our keynote, scheduled to be delivered by Beth Macy, author of *Dopesick* and *Raising Lazarus*. Beth was kind enough to come to campus in October to deliver her keynote belatedly, and it was the first event we hosted postflood. We cleared out the clotheslines of thawed and drying office documents just in time to set up chairs for her visit. Some of the writers were also among the first to return, for a multiday retreat postflood with the Southern Appalachian Writers Cooperative that fall. Being with writers their first time back after they fled that terrifying morning is beautiful and heavy, and the experience will continue over several years until all those who are going to trickle back to us do so.

In October, Kentucky governor Andy Beshear announced the flood-related death of a ninety-seven-year-old woman, who had gone viral in a photo of her sitting in bed, waist-deep in brown water, with her belongings floating around her. The governor's declaration of flood-related deaths still occurring months later shows the long-term view of such disasters: it is impossible to gauge the actual death toll because of the stress and health impacts on those who initially survived the event. There have been over forty known victims, who ranged in age from ninety-seven down to the youngest, a two-year-old who died along with three siblings.[5]

During the rescue phase and the beginning of recovery that summer, what struck me was how often I heard people say some equivalent of "We had six feet of water in the house. We lost the car and are now homeless. We lost absolutely everything, but we're OK because no one died." That is the bar for this tragedy, especially in Knott County, where twenty-two of those more than forty lives were lost. That is how bad it was. If you lost absolutely everything except your life, you were OK.

And nothing was untouched.

Like a memorial book trying to tell the story of a life, this collection zooms in to show us moments to help us process, remember, and try to understand the whole. I can tell you how my office was destroyed by fifty-two inches of water, and when Cassie Mullins Moses (daughter of former director Mike Mullins) was on campus with a crew in the immediate days that followed, I asked if they could flip my desk right side up so I could get my bank card out of a drawer. I also asked if they could help me find a small glass perfume bottle my dead father, a glassblower, had made. The filing cabinet it sat on was lying cattywampus over my upside-down desk, which was half bashed through what used to be a glass wall. When one of the volunteers in Cassie's crew lifted that unbroken glass bottle out of the mud and handed it to me, I absolutely broke down.

Someone had filled a bucket of creek water and left it outside for whoever needed it. There was no running water for days until Moses Owens, our maintenance foreman, was able to get our old, out-of-service well working. Before Moses got that working and several porta-potties were delivered, volunteers hauled up buckets of creek water a few times a day to flush toilets in the Mullins Center, as well as Stucky and Preece, where dozens of now-homeless community members were sheltering. I used one of those buckets to wash the mud from my father's glass. The next morning, preparing myself for the donations and people and emotional chaos, I took my thirteen-year-old, who was cleaning archive documents in the Great Hall, outside to the parking lot. There was glass sprinkled everywhere across campus, deposited by the water when it receded. My kid and I picked up a hundred little pieces, washed them in a bucket of creek water, and filled that unbroken perfume bottle with them.

It was a little nonword poem of the broken in the unbroken, in the hands of people both broken and not, in a community both broken and not. A little snapshot to humanize and individualize a community tragedy. To show others and ourselves a little of what it was like, the best we could. It sits on my piano at home—a home on the bank of the Middle Fork of the Kentucky River two counties over. If my section of the river were to rise twenty feet, my children's lives would be in danger, and I could lose everything, just like those affected by this flood. This is a reality I carry.

Through rescue and the beginning of the recovery, I can pull forth snapshots: my momma relayed important information from social media and news outlets to me so I wouldn't miss it; my husband packed the car each morning with coffee and supplies; our senior director, Josh Mullins, and other staff held Reading Corps training amid the chaos because we still were providing dyslexia and reading tutoring to schools that were in session and unaffected by the flood; our traditional arts director, Sarah Kate Morgan, in mud-splattered overalls, lovingly cleaned a historic Uncle Ed Thomas dulcimer;[6] a Dallas, Texas, police station called to ask how they could help; a Columbia Gas employee walked through the door, said he had an hour-long lunch break, and asked us to put him to work; the UPS truck off-loaded thirty, forty, fifty, or more boxes of donations each day; so many reporters came with cameras and microphones; community members struggled to sign up for, wait for, and deal with the Federal Emergency Management Agency (FEMA). A million miracles and heartaches.

Six months after the flood, I saw colleagues standing at the window by the dining hall, looking down to the creek to watch some community members pull a wrecked car from the water while stinging snow blew around them; the car had been carried a quarter of a mile, swept around the walk bridge, and deposited upside down on our section of Troublesome. Afterward, I went to sit in my makeshift cubicle in the back of the Great Hall to try to answer emails in this strange, new world.

A snapshot of recovery.

And that is what these pieces do. They offer strobe-light snapshots to say, "This is a little of what it was like."

And within that, we will see variations. In the dark, roaring early hours of July 28 before the sun rose and the waters receded, some people smelled gasoline. Some smelled propane. Some smelled diesel. Some people slept right through it and didn't smell anything. We will see things in jumbled order. As we read the collection, we will skip across chronology and then double back—memory is rarely restricted to a one-directional time line. As Ron Houchin wrote in his poem "To Make a Thing" from *Talking to Shadows*:

> I had no idea how
> my mind would work to fracture events,
> rebuilding them as memory.

Memory during a crisis can be hypercrystallized, or mushy blur, or jittery skitter.

This is also why I did not ask the writers to revise out all the points of repetition. Several writers will tell you how they awakened to find the terrible water, that they moved to higher ground, that they felt crushing guilt when they fled. This is a chorus that sets the unique verses together in a whole song. The repetition of the lightning, the displaced ducks, the mud, the dog barking in the bathroom, the roar and stink, and the calling out of extractive industry as an accomplice are a tethering chord progression.

When several of the writers and I got on a Zoom call that Saturday to offer and ask for help and support, to check in on one another, and to give updates on our friends who were still trying to make it home, we joked that this would be one of the best-documented floods ever since dozens of our best Appalachian writers had lived through it. We joked about when the submission deadline would be for an anthology (little could I have imagined I'd be its editor). We knew we would rise up and tell the tale with our fragments. All put together, it would show a little of what it was like. We'd put the parking lot glass shards into a bottle to tell a story.

Perhaps not surprisingly, I found many of the writers struggling to write about it. Some whose work I solicited hadn't written anything about it six months later. There were a handful of writers who initially said yes to the invitation to submit but then regretfully had to rescind. Or they jotted out the skeleton of a piece, but going back to those moments to put flesh and sinew on the bones was too difficult. Writing can definitely be therapeutic. But it can also be retraumatizing, and many folks just need more time.

But as a community, we did rise up to tell the tale. We are. We will continue to.

It seems an impossible task, to write about something this big. The flood itself seems impossible—that it happened at all. I stand on the walk bridge and see the quiet four-inch-deep trickle ten feet below me, and then I envision the water over my head, tearing through, and I can't understand it. Water with enough force to slam a five-hundred-gallon propane tank into the bridge and dent the iron cross braces, then carry the tank off somewhere downstream. Water with enough force to explode through our office doors, blast down the hallways, and still be strong enough to bend door hinges, crash washing machines sideways, and knock books from shelves. Water with enough force to carry a dumpster from somewhere upstream and deposit it on campus like it was nothing.

Because it *was* nothing really. Just physics.

Some of these pieces put a god or nature as villain or savior, but in the end, it was mathematical physics. Indisputably, humanity is a villain here because we greatly influenced these physics toward the deadly. Many of the pieces call out humanity's shortsighted choices that brought us to this point. And as in most human-made disasters, the people who had the least power over creating the situation suffered the most.

I had my sixteen-year-old with me at the AWW, helping with the trivia tournament on Wednesday night. We went home four hours before the flood hit. As we drove westward, away from my friends and mentors on campus, the rain sputtered, heavy to moderate. I have never seen as much lightning in my life. The sky to the north flashed constantly. We gasped, swore, and laughed about it. When the water rose, my family and I were quietly asleep forty miles away. I awoke inexplicably at 3:00 a.m. feeling queasy, and I looked at my phone, which was blowing up with alerts, texts, and calls. I experienced the flood itself on my phone, remaining warm, dry, and safe. But the pieces in this book teach me.

So much of my work at the Settlement School has changed because of the flood. We started the summer with our inaugural Ironwood

Writers Studio, a weeklong creative-writing camp for high schoolers, a historic addition to the Settlement School's literary legacy. I couldn't have known the summer would include FEMA officers and community members sitting at one table in the dining hall while at another UK Healthcare gave out hepatitis and tetanus vaccines for anyone who had been in the water. I couldn't have known that I'd be in my momma's mud boots, interviewing on TV about the archives and the needed supplies, or that my kids would end their summer volunteering well over two hundred hours each to help. I couldn't have known how much of my heart would change.

None of us could have known how different the world would be for us when the sun came up and the water went down.

Our writers will show you.

Through the process of pulling together this collection, I have metabolized a lot of my heartache, fear, and anger. As each piece came in over the winter and spring months, I dived back into the waters and came out a little altered each time. As I reread, a piece that hadn't made me cry now did. A piece that had choked me up for an hour on the previous read had more air in it.

In March 2023, FEMA reported that they had provided over $101 million in direct awards to survivors (to nearly 8,700 homeowners and renters), and the US Small Business Administration disbursed $58 million in disaster loans ($52 million in 724 loans to citizens and $5.7 million to 49 businesses and nonprofits).[7]

The Appalachian Citizens' Law Center and Ohio River Valley Institute estimated that it would cost $450–$950 million to rebuild the nine thousand homes destroyed or damaged by this flood. There is such a wide dollar range because to rebuild the homes where they were would cost $453 million and to rebuild the homes in *safer locations* would cost $957 million.[8]

People rebuilding in locations that flooded have been highly criticized by the outside media, but people from out of here fail to realize that there are few options. Yes, there is higher land, but is it for sale? Is it owned by some out-of-state corporation? Would it need to be cleared and leveled, or is the land ready to be built on? Are there roads and electricity up there? If there's no city water, can I dig a well up there? How far does that put me away from a school, grocery store, and job? Can I get cell phone service or broadband up there? Will the post office deliver to my house? These are not concerns to be flippant about. When my kids ride the bus to school, it takes forty-five minutes to get the nine miles there, and to this day the post office won't deliver to my house. The hospital, library, and auto-parts store are already a thirty-minute drive. If I had to move out of a floodplain, it'd be even worse.

In addition to all this, that report states that twenty-two thousand people were in homes damaged in this disaster, and six out of ten of those homes report annual incomes of less than $30,000. It is no secret that many of the areas affected by this flood experience financial poverty and its associated struggles.[9] Many businesses, schools, families, and communities were barely making it, and the security they had accumulated was suddenly gone.

And now a year out, recovery is chugging along. The Perry County FEMA trailer park I passed each day on my commute sprang up and was dismantled by spring. There are vehicles still stranded and filled with sediment in an adjacent parking lot. Some buildings on Main Street are completely repaired and open. Some are still boarded up. Many are somewhere in between.

On our campus, the dyslexia program is operating out of the second floor of the James Still Building. We repaired the front lobby, where students and parents enter before heading upstairs, but if you go through the door off the lobby, the hallway and other first-floor classrooms are still gutted. Repairs are being made to the ground floor of the Mullins Center and may or may not be completed by this summer's writers' workshop. We have raised some of the money to fix up Uncle Sol's Cabin. We are still operating out of temporary cubicles in the back of the Great Hall and tables in the Gathering Place. But our programming is vibrant and rolling. The second Ironwood Writers Studio of high schoolers from across the region, our Pick & Bow music program and after-school art clubs in the elementary schools, our Roots & Rows community gardening and farmers' market, dyslexia tutoring in schools throughout the region—for now it may look a little different than it did before, but it's all going strong. It is a well-repeated and deserving reminder that recovery is not a sprint. It is a marathon.

And this anthology shows that vital spirit of our institution and region.

Introduction

I want this collection to stand as a historic document for and about our writing community. Many of these pieces, while time-stamping the setting, say how transformative the AWW has been for them. As piece after piece says it, it begins to take on a mythic tone. But they aren't lying. It isn't propaganda. Some writers come one year, and that's it. Many, like me, find an immediate creative home and come back year after year, decade after decade, as often as they can. And many who call this gathering a home will express a terrible guilt about abandoning that home once the roads were passable and a terrible grief at being unable to be onsite to help during recovery. And our writing community extends beyond the AWW itself.

I want to honor that community and what they went through, whether they were on campus or not, because this community took me in after I moved here from the shores of Lake Erie. They took me in and became my creative soulmates. If I could've included all of them whom I respect and love, this book would've been as long as an encyclopedia set.

When looking at the table of contents, readers will see both very familiar names and some they have never seen before. The contributors range from writers earning their very first publication credits to those who have published literally dozens of books. And they are side by side without hierarchy or fanfare because the emerging writers in this book were just as stranded, just as threatened, and just as traumatized (or more so) as the more extensively published writers or those with a longer history with the AWW. The flood was an equalizing factor in its blind power. It didn't care who you were. And since we are so limited on space and there are so many living writers associated with Hindman Settlement School to choose from, I have not used space reprinting the work of our honored dead. Our famous writers like Albert Stewart, James Still, Jim Wayne Miller, Harriette Arnow, and others pivotal to the AWW can be read elsewhere. Their legacy is preserved already. I want this anthology to focus on the living and on what we have to say about this historic event we went through.

Additionally, I want this collection to show the wide angle of this situation. Climate change. The strength of community. The big, generous heart of the region. The need for change. And I want this collection to show the lacerating, individual human experiences.

So much held in a little glass bottle of parking lot glass shards.

In February the following year, we had a terrible night with storms across the region. The flooding was high, but a "normal" level of high ("normal" being a shifting concept). There was so much anxiety throughout southeastern Kentucky and the writing community. Folks didn't sleep. They piled up with their spouse and all their scared kids; they got up every hour to shine a flashlight on their waterway to gauge its height and threat; they kept scrolling social media for updates and alerts. When the body learns that a night storm is very dangerous, how many night storms until it learns that things will *probably* be OK this time? How long until the babies stop crying because the air conditioner sounds like heavy rain when it kicks on? How long until my mind, as I drive alongside a river, no longer calculates a twenty-foot rise to see who and what would be destroyed?

I hope this collection helps the community metabolize some of the trauma. I hope it shows the tremendous generosity and courage of ordinary people.

Our flood stories are unique, and also, taken in aggregate, they are not. When disaster hits, from tornado to drone strike, people rise up to save each other. In the end, all our differences fall away, and humans help each other. The people of this region, as they always have and always will, rose up to save each other during and after this terrible flood.

I want this collection to honor that—the searing, common humanity of the people of eastern Kentucky and the surrounding community who came to our aid in our dire need. It is a big story. And I want this collection to honor the part that happened to Hindman Settlement School, our writers, and our neighbors.

Melissa Helton
June 2023

Notes

1. For more information about Hindman Settlement School history, see Jess Stoddart, ed., *The Quare Women's Journals: May Stone & Katherine Pettit's Summers in the Kentucky Mountains and the Founding of the Hindman Settlement School* (Ashland, KY: Jesse Stuart Foundation, 1997) and Stoddart, *Challenge and Change in Appalachia: The Story of Hindman Settlement School* (Lexington:

University Press of Kentucky, 2002). For more information about Hindman Settlement School programming, see https://www.hindman.org.

2. The NOAA page for this event can be found at https://www.weather.gov/jkl/July2022Flooding.

3. "President Joseph R. Biden, Jr. Approves Kentucky Disaster Declaration," White House, July 29, 2022, https://www.whitehouse.gov/briefing-room/presidential-actions/2022/07/29/president-joseph-r-biden-jr-approves-kentucky-disaster-declaration-5/.

4. For more about the history of the Appalachian Writers' Workshop, see Leatha Kendrick and George Ella Lyon, eds., *Crossing Troublesome: 25 Years of the Appalachian Writers Workshop* (Hindman, KY: Hindman Settlement School, 2002).

5. Olivia Krauth, "Eastern Kentucky Flood Death Toll Climbs to 44, Beshear Says," *Louisville Courier Journal*, December 22, 2022, https://www.courier-journal.com/story/news/politics/2022/12/22/eastern-kentucky-flood-death-toll-climbs-to-44-beshear-says/69751648007/.

6. For more about Uncle Ed Thomas, Jethro Amburgey (whom our "new" bridge is named after), and mountain dulcimers, see Ralph Lee Smith, *The Story of the Dulcimer* (Knoxville: University of Tennessee Press, 2016); Dulcimer Museum, https://www.appalachianluthiery.org/museum-of-the-mountain-dulcimer; and "Uncle Ed Series Dulcimers," Troublesome Creek Stringed Instrument Co., https://troublesomecreekguitars.com/instruments/uncle-ed-series-dulcimers/.

7. For more FEMA information regarding this disaster, see https://www.fema.gov/disaster/4663. The March 9, 2023, report can be found at https://www.fema.gov/press-release/20230309/federal-disaster-assistance-eastern-kentucky-flood-survivors-tops-159.

8. Eric Dixon and Rebecca Shelton, *Housing Damage from the 2022 Kentucky Flood*, Ohio River Valley Institute and Appalachian Citizens' Law Center, February 21, 2023, https://ohiorivervalleyinstitute.org/wp-content/uploads/2023/02/Housing-Damage-from-KY-2022-Flood.pdf.

9. For Appalachian Regional Commission reports and data on population, economics, education, health, and more across the Appalachian region, see https://www.arc.gov and the graph-laden data book at https://www.arc.gov/news/arc-releases-new-data-revealing-appalachias-economic-improvements-key-vulnerabilities-compared-to-the-rest-of-the-u-s/.

Prologue

Sonja Livingston
Noah's Wife

Every living thing on the face of the earth was wiped out; men and animals and the creatures that move along the ground and the birds of the air were wiped from the earth. Only Noah was left, and those with him in the ark—his sons, Shem, Ham, and Japheth—together with his wife and the wives of his sons.

—Genesis 7:23

THE SKY IS GRAY. NO RAIN YET, THOUGH HAM'S WIFE FEELS something on her skin. "We'll know it when it comes," Noah's wife says, setting the girl back to weaving. She remembers the Euphrates rising when she was a child: waterlogged walls, a trio of roosters bloated and floating, silt-blackened courtyard trees. She touches her daughter-in-law's hand. "There's nothing subtle about a flood."

Noah's wife—what to call her? *Na'ameh*? *Emzara*, perhaps. *Norea, Haykêl, Set*? Mother of all that survived God's fury, yet no one thought to remember her name.

Noah's at the basin, washing his hands and singing prayers. "Soon," he says, "we move to the ark." His wife surveys the place. Not much to begin with; still, it does not fit. "No room!"—Noah's refrain these last six days. "Think of the elephants!" he said, plucking a red necklace and length of linen from her hands. "Wife," Noah said, "you must think less of things and more of God."

What does she know of God? Her life is chopping the heads off fish, keeping girls from tearing out fistfuls of hair when their time comes, laying out babies who arrive without the gift of breath. Sweet, shiny things. She hates to give them back to the earth.

She'd waited for God to tap her on the shoulder and explain how to house the water buffalo, which birds the sand cats prefer, how much wheat to grind. God was specific when it came to the ship—dictating dimensions and type of wood—surely, he'd advise on quantities of lentils and soft cheese, ways to survive forty days and nights of mewling and animal mess.

But God did not come. The women did what they could: dried strips of fruit and meat, plucked chickens till their hands bled, loaded so much bread onto the boat, the barley scent overtook sawdust. They rendered sheep fat with ashes for soap, infused oil with juniper, hid cedar and olive saplings between beams, sewed seeds into the hems of their robes. They shook pistachios from branches, gathered pomegranates, traded for dates—for the flood, of course, but also enough for a cake on Noah's birthday because, end times or not, a man does not like to be forgotten.

They're loading the last of the animals now. Noah's wife is in the garden, trying to memorize each frothy leaf and turn of vine. From the neighbor's house, the crackle of fire and scent of meat. Overhead, a leaden cloud. She leans against the palm tree, watching the stars disappear one at a time. How long does she stand there looking up? Five minutes? An hour? There's no such thing as time. Until Ham's wife rushes over, beads of water glinting on her forehead. At her feet, the soil's pockmarked and wet.

Electricity, sharp and white, as they load themselves onto the boat. A sob from Japheth's wife. Sudden bouts of clarity concerning what they can't live without. Noah's wife slips the beads her husband forbade around her ankle. Carnelian, from her mother. Noah's no spring chicken—and neither is she—but she's built like a reed, and nights Noah gets into the wine, he says she still looks like a girl. He calls her name through the rain. "Woman," Noah shouts. "Come."

She will come, of course; what else can she do? But first she runs. Her robes are soaked when she reaches her sister's hut. Everyone's laughing as the rain chases them inside. A boy slaps his feet against the courtyard mud as the fire spits the last of its flames. Noah's wife pushes through the crowd, touching every face. "Be kind to the little one," she says when she reaches her sister.

"That boy," her sister says, cursing the child still playing in the rain. "More trouble than he's worth."

Noah's wife sets a finger against her sister's lips, buries her face in the mantle of black hair. Outside, Noah shouts his wife's name into the wind. Outside, little rivers begin to form.

"Let that baby sleep beside you." Noah's wife turns away. "Just this one night."

Downtown Hindman. © Britton Patrick Morgan

The Map Keeps Changing

Jesse Graves
Hanktum

I have asked around to see how far the word
traveled, and never found it familiar
to anyone more than a few hollers removed
from Cain Road and the Katie Myers ridge.

So much of the family lexicon died
with my aunt June, her way of calling the cows
home from the fields, her names for the herbs
and berries that went into medicines and cakes.

What do people call a fearsome storm, one that
moves in from the northwest and builds
in layers of darkening clouds, if not a *hanktum*?
Lightning strikes, and none can say its true name.

Julia Watts
After This, the Deluge

THE FIRST THING I DID WHEN I ARRIVED FOR MY ONE-WEEK summer teaching gig at Hindman Settlement School was lock myself out of my room. My partner had driven me from Knoxville, and we had found my key in its designated hiding place, opened the door to my sweet little apartment (decorated in spring green, with a huge painting of a benevolent-looking cow hanging next to the bed), and unloaded my week's worth of clothes and groceries. We had brought in everything except my water bottle, which I had left in the car's passenger-side cup holder. I went out to grab it before my partner headed back to Knoxville, and I let the apartment's door shut behind me. I didn't know that the door locked automatically.

It was a Sunday afternoon, and the campus was deserted. Did I have contact information for staff who could help me? Sure, I did. On my phone, which was locked in the room with my key and what at the time felt like all my worldly possessions. I searched the campus in the hope of finding someone from maintenance who might have a full set of keys, but to no avail. Finally, out of desperation, we drove to downtown Hindman, where we found a small redbrick building that identified itself as the Volunteer Fire Department. A young woman answered the door. I asked if she worked for the fire department, and she said no, that she was a girlfriend of one of the guys, but that we could come on in. The four guys—all of them well under thirty—were sitting on a couple of dilapidated couches, playing on their phones and eating chips. Bottles of Mountain

Dew were close at hand. The station could've been a single guy's dumpy apartment.

Consumed with embarrassment at my incompetence, I said, "Hi. I'm a guest teacher over at the school, and I did a dumb thing."

Once I explained, they swung into action. "Give us five minutes to get our stuff together, and we'll see what we can do," the tallest of the young men said in a reassuring eastern Kentucky twang. The guys then went into full MacGyver mode, grabbing axes, a giant key-ring-looking thing, a crowbar. They worked with good cheer, seeming happy to have a project to break up a slow Sunday afternoon. Even though it seemed like overkill, they all followed us back to the school, forming a small caravan of pickup trucks.

They worked for about an hour while I grew increasingly anxious. They called a guy from the larger fire department in Hazard, who had a tool they lacked. He showed up about twenty minutes later, and whatever magical tools they keep in Hazard seem to be remarkably effective. When the door burst open, there were high fives and hallelujahs. I thanked all the young men profusely, and once they were gone, I laughed—both at my own stupidity and at the fact that I had given bored small-town volunteer firefighters something to do on a sleepy Sunday. Knowing what I know now, my attitude seems condescending, with me thinking that my moment of absentmindedness passed for an emergency in Hindman. How could I have known that only a month later, these firefighters and other emergency responders would be working around the clock to help the victims of a flood that had caused an unthinkable loss of life and property?

My gig was teaching a creative writing class in the Settlement School's excellent program for kids with dyslexia.[1] While I have decades of experience teaching creative writing to students of various ages, I had no experience teaching students with dyslexia. Thankfully, the three experienced staff members in my classroom could step in to give the kids one-on-one help and deal with any disciplinary problems that might come up. I was to teach three different sections of the class, starting at 8:00 a.m. and ending at lunchtime. The kids were great: friendly, funny, and energetic, with eastern Kentucky accents that made me feel right at home. They were at that age where the boys ragged on each other good-naturedly but constantly and sat together at a table as far as possible from the girls, as if being female were a communicable disease. The girls talked to one another more quietly, often in pairs. The boys, for all their raucousness, might as well have been invisible to them.

I asked the kids to introduce themselves, talk about what they liked to do for fun, and tell the class about their pets. Having been raised not too far away, I figured these kids were growing up surrounded by animals like I did. They talked about their favorite video games and the sports they played. One little boy with a 1950s-style flattop offered that he had "a big ole fat dog named Buford." A girl with long blond hair and freckles announced that before coming to class that morning, she had been bitten by her chicken. The ice was officially broken.

Since the kids struggled with reading and writing, I tried to create prompts and activities that would maximize their use of imagination but work for a wide range of skill levels. One was, "You are the ruler of a kingdom. Describe what it looks like and what the rules are. Are you a fair and kind ruler or an unfair and cruel ruler?" Another: "You are a dragon. Describe what you look like and what your powers are." For the record, all the boys chose to be unfair and cruel rulers, and all the girls chose to be kind and fair. The girls were all pretty, nice dragons, and the boys were all terrifying ones.

When given a prompt, many of the kids asked if they could draw pictures instead of writing. I figured this made sense—if you struggle with reading words, it's logical that you would think in pictures. I told them that I was fine with them drawing but that they also had to write at least a few sentences, that their drawings should be used to illustrate their writing. I proceeded to hand out colored pencils and crayons.

The kids generally flew through the assignments and used the remaining free time to draw Pokémon, Minecraft dragons and zombies, and an apparently wildly popular blue monster named Huggy Wuggy that I had never heard of and had to Google. For some reason—maybe because most of the kids were really young or just because it had been a long morning—the last class before lunch was always the most chaotic. One little girl in that class (I'll call her May) was particularly memorable. Some of it was her sheer cuteness—seven years old, built like a little teapot, with curly brown hair and glasses that made her already-big blue eyes look huge—but much of it was the force of her personality.

On the first day of class, we played Liar's Club, a game in which everybody picks a random item out of a bag and has to make up a story about it. May reached into the bag and pulled out a good-luck kitty figurine, gasped with delight, and said, "Oh, is this a gift for *me*?" I explained that, no, the kitty figurine actually belonged to my son but that she was borrowing it so she could write a story about it. She was silent for a moment, then said, "I have a great idea. I think we should have a day in class called Gift Day where everybody brings me gifts." I told her I'd take it into consideration.

May tried to draw pictures during the excessive amount of free time this class always seemed to have, but she was never happy with the results. "It has to be perfect," she'd say, and I'd try to explain the dangers of perfectionism in language a seven-year-old could understand. Somewhere in this process, she discovered that I can draw. I'm not great, but I'm pretty good by elementary school standards (if I draw a dog, it looks like a dog). And so, over and over again, May would thrust a sheet of paper at me and demand a drawing of whatever was in her imagination: a sleeping kitty, a baby deer, a unicorn with a rainbow coming out of its horn. I was turning into a cut-rate Lisa Frank. Each time I completed a drawing, I'd hope she'd be satisfied so I could move on to something that approximated teaching, but she'd always look at my handiwork, nod, then hold out the paper again and say, "Now draw me a different picture."

To be honest, I'm not sure the kids learned a damn thing under my tutelage. I hope that, at the very least, they learned that written words, which are often a source of struggle for them, can also be a source of creativity and imagination. And I hope they had fun. I certainly did.

I wasn't there when the floods came. Several of my friends were, attending the annual writers' workshop as either faculty or students. As they waited for rescue, they posted pictures of the place I had been just a few weeks before. The images defied belief. Troublesome Creek, which had not even been ankle deep during my residency, had swollen to completely submerge the bridge I had walked over every day to get to class. One friend, a workshop faculty member, had been staying in the same little green apartment I had locked myself out of. She posted a picture in which the benevolent cow in the painting seemed to be surveying the floating wreckage that remained of the room. I thought of the cultural touchstones I had seen during my time at Hindman: the school's outstanding collection of Appalachian books, now washed away or destroyed; the dulcimers in Hindman's downtown dulcimer museum, gone.

Of course, as the news poured in, it was much worse than stranded friends and property damage. Deaths of mothers, fathers, grandparents, and children. Loss of wildlife, farm animals, pets. Destruction of homes and farms. The flood-relief efforts on the local, regional, and national levels were truly heroic, but with a writer's obsessiveness over words, I found myself getting hung up on the word *relief* in *flood relief*. Certainly, the government, the Red Cross, and the army of volunteers, which probably included the Hindman firefighters who had helped me with my

Flooded cars with emergency lights on. © Britton Patrick Morgan

trivial problem, provided much-needed assistance. But relief? Some pain can't be relieved.

There are many writers, scientists, and scholars who are far more qualified than I am to explain the various environmental and socioeconomic factors that made the eastern Kentucky floods such an unmitigated disaster for the people of the region. But in the aftermath, I've just found myself thinking about the terrific kids I taught last summer and wanting things to be better and easier for them. I wish I could draw them a different picture.

Note

1. The dyslexia education program began in 1979 as an after-school, parent-led tutoring program. The program has evolved over the years to serve more than one thousand children and families annually through daily intervention services at twenty partner school sites across the region, weekly on-campus and online after-school tutoring, and an intensive five-week summer program. Learn more at https://hindman.org/dyslexia.

Bernard Clay
they freaks of nature

they say it was a thousand-year flood
that hit the cumberland plateau
and turned creeks into niles
and rivers into caspian seas

this place
that eons of tectonic collisions,
glaciation, and erosion
have architected
into a perfect rain-percolating
nursery for birthing headwaters
this is the place?
that can be flooded?

this same "they" will also say
no humans inhabited kentucky
a thousand years ago
because "they" only detect people
by what has been destroyed
and left behind
but the ones here back then
were so integrated
into the robust ecosystem
(that "they"

later methodically dismantled)
there was hardly a trace left
by the time real science
was discovered

but if asked back then
those indigenous folks
probably with confused
what-are-"they"-talking-about? faces
would say "it's water
how can there be too much?"
while now
in the news every other day
a thousand-year flood pops up
in a new spot on the planet

heck if the appalachians were asked
if floods like this had been seen back
ten thousand or even ten million years ago
those mountains
would chuckle and say
"we haven't seen anything like this
since we were ocean bottom
during the devonian age"

Maurice Manning
Blue Hole

There are places in streams and little rivers
where the current slows and the water is deeper
and someone decided to call these holes.
Well, that's the local name I know.
A hole in a little river or stream,
a hole you can see in water. Somewhere
in a photograph I have, my father's
father is holding a stringer of fish.
It's late in the 1930s I guess,
my father as a boy is standing
on a splay of rocks above a shoal.
There must be thirty fish on the stringer.
They'd been to the hole that day. In fact,
they called it a blue hole, because
the water was bluish green. It was down
on Redbird. I like the grammar
of that particular phrase. It was down
on Redbird, that day back then
when they found a blue hole and the hole
created by Time was full of fish.
Down on Redbird. The water
over your head but you could see
right down to the bottom of the hole.
A river named for an Indian

supposedly kind to settlers.
That's the legend. You can also see
in the background some scraggly trees.
The bank looks low, an easy way
to get to the water. Nobody
in the picture knew what was going
to happen, but another kind
of hole was coming. It wasn't blue,
and it would swallow everyone
standing in the photograph taken
down on Redbird that day,
a hole not made by Time, but by
the son of man, living in Time
and lost in the blur of Time and grief
and no one was coming back from it,
because it doesn't have a bottom.
This is a meditation on language,
geography, and something else,
something that doesn't have a name.
By son of man, I mean the man
who's stretching the stringer of fish before him
like a grin, whose voice I never heard,
though I was always told he could sing.

Leatha Kendrick
Invisible, Essential

FIRST, THERE IS PARALYZING GRIEF—A DEAFENING, IF SOUNDLESS, howl that wants everything back as it was—wants families waking up in their beds, grandmothers safe in their morning kitchens, and homes intact and dry. No one dead, no wall of water tearing down narrow valleys between hills stripped of trees and soil, no record rainfall coming faster than the creeks can carry it away, no mud and debris, no washed-out bridges. It's late in the afternoon on July 28, 2022, and from the distance of my writing room in Lexington, I am listening to George Ella Lyon on the phone as she narrates how she fled from the ground-floor apartment up the stairs in the Mike Mullins Center and then up the hill through darkness and rain to Stucky; how she and the others at the Appalachian Writers' Workshop who'd been trapped at the Hindman Settlement School finally (agonizing hours later) managed to cross the mud-covered, treacherous bridge; how they caravanned along a circuitous route of detours around flooded roads and away from the devastation. I am numb as I realize how narrow her escape was. At first, because I did not have to move through mud and wreckage, I reject the reality of loss.

  I drove across a flat wooden bridge onto the Settlement School campus for the first time in August 1987, expecting to spend two or three days commuting to the Appalachian Writers' Workshop. George Ella had urged me to attend. I had three young girls at home near Prestonsburg, about forty-five minutes away. I thought it would be easy to come and go, sit in on classes, and maybe hear one of the evening readings. By the second day's lunch, I did not want to leave. Like many people who have come to the Settlement School over the years, I will return again and again—as a writer, and later on as a workshop leader and editor, embraced by a close-knit, if far-flung, community.

Back then, before Google, before Garmin even (remember Garmin?), we had to find our way to Hindman somehow if we wanted to go to the writers' workshop. Someone had to chart the way and pass the map along. Because the Settlement School lay far off the beaten path, the back of the mimeographed, trifold writers' workshop brochure included a crudely drawn map. Floating in the center of the page, it could have been a rendering of the main lines in an open palm, with Hindman roughly at the heart of its intersections. The artist had included only the main highways, leaving out extraneous roads and place names. The essential ones— the Mountain Parkway, Route 80, Route 23, Route 119—were darkly inked. "Heavy lines indicate best route," someone (probably Mike Mullins) had typed onto the map's edge. It is a map we could still follow today if something wiped out the internet and all our GPS devices failed.

In 2002, when George Ella Lyon and I conceived and edited *Crossing Troublesome: Twenty-Five Years of the Appalachian Writers Workshop*, that map went onto the title page. A hand-drawn rendering of the campus occupied a double spread on the next pages. It felt essential to capture the physical presence of the Settlement School at that moment because everything was poised to change. A new library/learning center would soon replace the old Knott County Library. The railless wooden bridge crossing Troublesome Creek at the campus was set to be superseded by a modern concrete bridge. In the years to come, all of the buildings on campus would be renovated, some nearly beyond recognition. I wanted to hold that world still, so I drew an illustrated map of the campus as I knew it. My model was the Hundred Acre Wood map at the beginning of the Winnie the Pooh books—drawn with trees and abodes and hiding places. A magical landscape.

When I look back at that map now, I mostly see what it cannot capture: the feel of the workshop. Those of us who've been to the Appalachian Writers' Workshop know the campus as an embodied place, a place apart, with its own weather, its own rituals and traditions, and, yes, its particular configuration of buildings and landscape that exist both within and outside time. For one intense week each year, the workshop rises there. Our attention sharpened through hours of talking and listening, we receive the intense sensory experience of Hindman in full summer (the dense air, heavy with moisture; the sounds of insects and birds; the rush of traffic echoing off the Carr Fork Road cut-through; the steam of the industrial dishwasher and the clatter of plates and silverware as we clear and wash up after meals; the rise and fall of voices; the songs and bursts of laughter echoing from porches). Though the workshop's writers disperse to the four winds each year, we are changed and enlarged by discoveries made and doubts overcome. We have become part of something large and ongoing.

What is essential but invisible to the eyes (to paraphrase Saint-Exupéry) is the spirit embodied here—a vibrant scholarship, a living literature, and a generosity alive in what Mike Mullins once called "one of the most supportive environments you will ever find." What mattered about Hindman was not something I could map, though my loving depiction of the place served as a memento of what I had found there.

If the Settlement School's spot of ground seems to contain a power out of proportion to its remote location and humble facilities, it's because that power is real: this place at the Forks of Troublesome has served as a portal to larger worlds since the school's founding in 1902. Katherine Pettit and May Stone arrived with a determination to create opportunities for education and better health. But it was never a given that anything like the Settlement School would survive on this bit of earth. In the early days, the school suffered repeated setbacks; fires especially plagued them, in 1905, 1906, and 1910. The first large facility built at the Settlement School, the 1905 Loghouse, did not survive long. "On November 10—less than two months after its completion—it burned to the ground," Jess Stoddard writes in *Challenge and Change in Appalachia: The Story of the Hindman Settlement School*. "In less than twenty minutes the building and its contents were gone. The survivors stood there on the frozen ground in bare feet, grateful that the thirty-four residents were safe. When day broke, they walked down to the store to get underwear, shoes, and stockings. . . . Not only were all of the personal possessions of the teachers gone but the library of over two thousand books had been destroyed as well."

The planning, design, labor, and craftsmanship embodied in the building, the books painstakingly assembled—all gone. But rather than settlement at Sassafras twenty years later. Her reaction to the damage she saw from the coal mining along the creeks of the area was so strong that she could not continue the exploration of her old haunts after the first day because she felt so depressed at what had happened to this formerly beautiful and remote corner of the mountains."

As I write this essay, everything I have read or seen feels like part of it, filled with meaning, in that way that happens when we grope for language to do justice to our dearest things. In *Lost & Found*, a wonderful memoir by Kathryn Schulz that grapples with grief and joy and the things that allow us to go on, I read, "It is easy to feel small and powerless; easy, too, to feel amazed and fortunate to be here. On the whole, though, I take the side of amazement. . . . For now, at least, the world is ours to notice and to change, and that seems to me sufficient."

"Ours to notice and to change." We are caretakers, our role here temporary, as Schulz also points out at the end of her book: "None of us would be here without what came before us, and none of us can know how much and in what ways everything that will come after us depends upon our being here."

When lives are lost and entire towns wiped away by the winds and fires and floods born of a new climate reality, it occasions a grief so profound as to be paralyzing. And yet this *now* is our time. It is all we have—all we can use to honor the people who have gone before us and the love that brought our lost worlds into being.

The map keeps changing, but what it maps—invisible, essential—remains.

Christopher McCurry
It's Raining This Week

And has been raining
for some weeks now.

Not every day, not
all day long. Just

consistently enough
to remind me of you

agreeing with me.
Sometimes the world

feels made for us. But
you make the world.

Storm God, Bringer of
Rain, your sadness soaks

the earth, seeps into
the dirt, the crawl spaces,

floods garages, picks
up houses and breaks

them into debris. Sweeps
what's left out to sea.

No one ever thinks
to ask a god how it feels

to bury her father. Would
you believe there's

sunshine and thunder on
the same day, that gutters

choke they're so full.
And the anger, the rage,

like a baseball bat shattering
pretty white teeth. It
sounds like glass, like cold

hard water pelting a metal
roof. We are fools.

We worship a malefic god.
She asks us for love

and we give her a world
full of grief.

Annette Saunooke Clapsaddle
Flood Walking

My friend, Mandi, tells me I have a phenomenal flood walk. OK. Maybe she didn't quite say "phenomenal," but close. When I crossed a Knoxville street amid bone-chilling rain to get closer to the stage door that we hoped recording artist Jason Isbell would exit after his concert, Mandi laughed and accused me of invoking the confident stride for the purpose of an autograph.

She's not wrong. I do have a flood walk—likely one I perfected during my days riding mountain bike trails, when I had to remove large limbs from my path. As a high school teacher, I also earned the reputation for being the first at the site of any teenage brawl in the hallways. A purposeful walk will get you to the thick of things rather quickly, removing obstacles in the process.

We can laugh about this particular trait (or skill, depending on how you look at it) now, but I still wish that it were something Mandi never had to know about me—that she never had to see me perform a flood walk.

July 28, 2022, the morning following one of eastern Kentucky's most devastating floods, is one of those unforgettable times that attach themselves to us like lichen, searching for the sunlight but never fully able to free themselves from the darkness. Even writing about this date feels like a struggle toward illumination that is just too far out of reach.

July 28, 2022, began over ten thousand years ago. July 28, 2022, began over three hundred million years ago.

The Appalachian Mountains are what connect me to Kentucky, what connect me to friends like Mandi. They are ancient and vastly different as they span the states they cover. And for me, they will always be Cherokee.[1] Regardless of the borders, barriers, and property lines that erupt across the range, they are a Cherokee responsibility. Property ownership, essentially authority over nature, is a Euro-American concept that has done little to safeguard the most important assets of this continent.

As a citizen of the Eastern Band of Cherokee Indians (EBCI) and as someone who was born, was raised, and lives in Cherokee, North Carolina, I am quite firmly tied to this place. I in no way mean this in a restrictive sense but see it more in terms of worldview. As the first published-novelist citizen of my tribe, I am often asked to be a representative of both the EBCI and Indigenous Appalachia as a whole. I am comfortable in this space, though quite adamant about reminding readers or listeners that I represent only one perspective, one experience. I consider myself a voice *of* this place, not *for* this place.

So, a few days before a speaking engagement in Virginia that would offer a platform for this voice, I drove onto the Hindman Settlement School campus. This is an environment that always reminds me of the complexity of Appalachia—a reminder that I knew would serve my visit to Virginia.

Hindman is a magical place for writers. As I turned onto the parking lot, I slowed to allow literary legend George Ella Lyon to pass with a welcoming wave. Scenes like this are common during the Appalachian Writers' Workshop (AWW) held each July. Some of the greatest Appalachian (and, I would argue, American) writers have strolled the grounds of the Settlement School, side by side with fledgling writers eager to learn. I was there to visit friends, meet with my publisher, and take in a few of the evening AWW offerings before traveling on to Virginia. It was going to be a great few days.

And it was a great few days. By Wednesday evening, my belly was full of homemade tomato pie, my future had been charted by a friend's annual tarot card reading, and I was inspired to tackle my next writing project. I talked with fellow writers about how to have hard conversations regarding diversity, especially with audiences who are often homogenous.

Wednesday night brought rain, subtle enough to admire from the porch of the May Stone building. It gradually grew so heavy that I marked the culverts filling as I drove up the mountain to the house I was sharing with two ornery cats and two far less ornery writers (including Mandi) for the week—just across the right fork of Troublesome Creek and

Highway 160. The rain was soothing, and I quickly crawled into bed and fell into a deep sleep.

I sleep with noise-canceling Bose earbuds. I turn off all notifications (with the exception of those from immediate family members). I wear a sleep mask. Yes, I am over the age of forty and am fine with the optics. This coma-seeking sleep has done nothing but benefit me in the past, but on the night of Wednesday, July 27, 2022, and the morning of Thursday, July 28, 2022, this practice became very dangerous.

On Thursday morning, as a dim light cast a grayness over the town of Hindman, my roommate Rachel prodded me awake. When I sat up and removed my earbuds, she said to me, "There is an emergency." All I smelled was gasoline.

Most have seen the pictures of the devastation of that night's flash flood in eastern Kentucky. It was the first time in my life that I had witnessed such.

Flooding at home has sometimes breached bridges and covered the floors of downtown businesses. It has caused slides and temporarily rerouted roads. Still, I have never seen such an immediate and total loss of land, property, and life.

I instantly realized that any judgment calls I made on whether to leave or stay, drive or walk that morning would be uninformed. I was not of this place. I did not know how these mountains would respond or when the creeks would recede. I wanted to leave and find the familiar before the asphalt beneath my car completely washed away. If I were home, I'd at least know which turns to make. Which roads might be passable. How much strain a riverbank could take.

And in the following moments, I realized the locals knew the way. They knew the patience and the urgency of each decision. They knew acceptance and resistance. They knew what was truly important, beyond lives, to save.

Mud-covered dulcimers and guitars lay drying on the floor of an auditorium that typically plays host to world-renowned-author readings. The printed works of these literary giants were sought out and recovered like buried treasure amid sewage-laden rubble. There were constant informal roll calls of sorts, ensuring that everyone was accounted for. The staff immediately began working to help others or recover property even though their homes had just been destroyed.

In fact, that concern for others and for the irreplaceable heirlooms that symbolize this place is what struck me the hardest as I drove away from Hindman for a second time, the first having been diverted by floodwaters and emergency vehicles. When I stopped back by the campus to get directions because GPS and cell service were nonexistent, I picked up a college student eager to connect with their family in Virginia. This was their first experience at AWW. As we drove, still unsure we were headed to safety, my phone kept ringing. Text messages kept coming in. Each one was displayed on my car's screen, so my passenger was privy to them all. When I recall today the names lit up on the monitor and the voices coming across my phone's speaker, it makes me teary eyed. Writers from across Appalachia and North Carolina were the first to call. The first to text. They did not hesitate to know everything. And soon they would answer the call to help remove the obstacles.

I could have taken pictures showing the extent of the wreckage and posted them to social media, but this felt wrong—voyeuristic. I couldn't look at my eastern Kentucky friends, like Mandi, and invite them to share their story in the high ground of North Carolina. Pimp their pain, so to speak.

They were already doing the work they needed to do. If I or others wanted to help, we would listen to their advice, return to our own communities carrying their stories, and be poised to move in the directions they would choose or need in the coming days, weeks, and, it now appears, years.

The guilt that lingers reminds me that I got to go home. I was able to leave and regroup. I escaped the nauseous gas fumes and unstable bridges. I drove to clean drinking water and dry clothes. I knew I would return soon, but I had a choice. So many of my friends and their loved ones never had a choice. My phone rang with my concerned friends and family, but those people were also safe. I wasn't tasked with locating my children or my father or my neighbors.

Back home in western North Carolina, I experienced a sense of disorientation. A few days after my return, my son attempted to wake me from a deep sleep. I panicked, scaring him half to death. Every time I drove by our rivers and creeks, tucked safely within their banks, I became angry. And months later, this hasn't changed.

When I see power lines at home, I think back to postflood recovery trips I took to eastern Kentucky and remember their power lines, those that somehow survived, marking the walls of water that came rushing off coal-scarred mountains. Those lines are still decorated with mattresses and other debris like tinsel on Christmas trees long past Epiphany.

When I try to describe to friends back home the loss I witnessed, I often say, "This wouldn't happen here." I want them to know that they likely cannot connect the magnitude of devastation to anything they have known firsthand, though those who have lived through hurricanes in places like Florida seem to more readily empathize.

But really when I say "this wouldn't happen here," I mean that I know it could. I am trying to convince myself that we have protected our portion of these sacred mountains from the ravages of big business. Cherokees are still here to steward.

However, that's me just retreating home again. Retreating to the familiar and the safe. That is me absolutely not getting into the thick of things. In reality, western North Carolina is one more luxury housing development away from a mountaintop-removal-induced catastrophe. The coal we burn, the garbage we toss, the pollution we create does not stop at the state line or even the border of the Qualla Boundary.

Cherokees may have been stripped of titles on a map, and those who erased them with abandon may deserve blame, but our culture still reminds us that we have a responsibility to these mountains.

I was fortunate enough to be able to return to eastern Kentucky within a couple of weeks. It was a gift of humility and an honor to work alongside the people who call it home. I joked that I wanted to become a Kentucky Colonel after spending so much time in the state—although the fine folks of Hazard reminded me it is much better to be a duchess of their town.

What they may not know is how much anger they relieved me of during those workdays. While I once thought of the destruction of the Cherokee mountains by outsiders, I now saw that stewardship was also capable of crossing borders and boundaries. The people of eastern Kentucky gave me hope when, as volunteers, we thought it was our job to inspire.

It wasn't the pallets of water or mounds of used clothes that poured in from across the country (including some rather sparkly, formal, three-inch-high heels) that helped to bring this region back from the brink of erasure. It was the people of this place. It continues to be the people of this place.

View of flooded Hindman from McLain Chapel. © Amy Le Ann Richardson

These Appalachian Mountains will always be a Cherokee place. No land deed will change that. No declaration of mineral rights will change that. No late-night storm will change that. These mountains will always be a Cherokee place, and I will always find it a privilege to flood walk my way across them with eastern Kentucky friends.

Note

1. Southern Appalachia is Cherokee land. In the 1830s, the federal government attempted to forcefully remove Cherokees in the Southeast. More than sixteen thousand native people were marched on what would historically become known as the Trail of Tears and relocated to Oklahoma. Between 25 and 50 percent of the Cherokee tribe died on the Trail of Tears. The Eastern Band of Cherokee Indians are those whose ancestors resisted removal through a series of strategic and courageous acts. These ancestors purchased fifty-seven thousand acres of property. This land, called the Qualla Boundary, is owned by the Eastern Band of Cherokee Indians and kept in trust by the federal government.

Patricia L. Hudson
Forty Years and a Flood

IT WAS RAINING HARD AS MY HUSBAND AND I DROVE OVER THE bridge above Troublesome Creek and onto the campus. I was scheduled to teach an afternoon session on historical fiction at the 2022 Appalachian Writers' Workshop, but I'd first set foot on the campus exactly forty years earlier, a fledgling freelance writer assigned to interview Appalachian author Harriette Arnow, who was a faculty member during the workshop's earliest years. After interviewing Arnow in her apartment in the main campus building, I'd walked around the grounds and in one afternoon discerned that the settlement was a magical place. As a writer, I felt as if all the words that had ever been read or taught or uttered in that space continued to hang in the air, so that writers who ventured there could breathe them in and come away inspired.

In the years since then, I'd returned as a workshop participant as often as I could, spending an always-too-short week at the Forks of Troublesome, communing with other writers and drawing energy from the gathering. It had become a ritual of sorts for me to pause each day on the iron footbridge that crossed the creek and watch the gentle flow of the water below. The highest I'd seen Troublesome on most of those summer days was barely more than a foot deep, with the water so clear you could see the stones along the bottom. For forty years, Hindman was a place I returned to again and again, a touchstone for my writing life.

However, in the spring of 2006 I'd arrived on campus for an entirely different reason—not to write but to listen to area residents whose lives had become intolerable because of an exceptionally destructive type of coal mining called mountaintop removal. The Hindman Settlement School was hosting mountaintop-removal listening tours, organized by Kentuckians for the Commonwealth. I'd been invited to attend as the cofounder of a Tennessee-based group called LEAF (the Lindquist Environmental Appalachian Fellowship), which was working among church communities to raise awareness of this imminent threat to the Appalachian Mountains and all the reasons it needed to be banned.

The group that gathered on campus came from all across Appalachia. We carpooled from the settlement to the foot of a nearby ridgeline, then hiked upward through the cool green of a hardwood forest where flame azaleas were blooming amid the understory. When we reached the top of the ridge, we walked along the crest until the trees ended abruptly, and we suddenly found ourselves standing on a barren plateau. In front of us was empty air—the ridge had been sheared off as if by a giant axe, and below us, in every direction, there was nothing but a gray moonscape. No green. Nothing left alive. Just rubble and rock.

We stood there, stunned, some of us in tears, and before we'd caught our breath, from somewhere below us a warning siren blared. Moments later, an echoing blast shook the ground, and we watched in horror as part of the ridge across from us was blown high into the air, vanishing before our eyes. The anguish I felt that day as I witnessed the death of a mountain has never left me. I couldn't fathom the hubris it took for humans to destroy a mountain that had stood, solid and full of life, since the last Ice Age had carved Appalachia's peaks and valleys.

That evening, back at the Settlement School, we sat in the Great Hall and listened to the stories of community members who spoke of black water running from their taps, of the expensive necessity of buying bottled water to drink, or even to bathe in, because their well water was contaminated with high levels of arsenic. They spoke of the fear they felt whenever it rained, as the runoff from the remaining hillsides brought cascades of mud into their yards and sometimes into their homes.

After that day, anytime I drove to Hindman, I recalled their words—their sense of helplessness—as they'd spoken of seeing their communities devastated, of being forced to suffer while people from elsewhere profited. During each trip, I'd noted that the signs of mining, which had once been largely hidden from view, had crept closer to the edge of the Hal Rogers Parkway, the main road into the region. Every year the damage became a bit more visible.

Now, in 2022, the rain was pounding down as my husband, Sam, parked our car on campus, and I unloaded the materials I intended to share with my class. Normally I would have stayed the entire week, but the risk of taking COVID home to an immunocompromised family member meant I'd arranged to teach my class, visit with friends, and then drive back home.

We'd arrived just an hour or so before my lecture, and a strong wind was blowing the rain sideways as we hurried to the rear of the Mullins Center and knocked on the door of one of the ground-floor apartments. George Ella Lyon, a longtime friend who knew how COVID-phobic we were, had invited us to eat lunch with her in her room, knowing we wouldn't feel comfortable removing our masks in the dining hall.

When George Ella opened the apartment door, I was momentarily struck dumb. I'd heard that the entire first floor had been renovated since I'd last seen it, but the spartan accommodations I remembered from years past had been totally transformed. It was like stepping into a posh bed-and-breakfast, complete with an upscale coffee station by the front door and color-coordinated furniture. It was lovely but disorienting, as if my childhood summer camp had magically become a stylish resort.

My interview with Harriette Arnow had been in this space, and several times over the years I'd roomed with George Ella here, an area that was generally reserved for faculty because the apartments were one of the few places on campus with private bathrooms. Those bathrooms had been the only bit of luxury before the renovations, as the space had still sported bunk beds and other pieces of hand-me-down furniture. One summer I'd pulled a mattress off the top bunk and slept on the floor all week, having reached the age where my knees complained about climbing up and down.

Sam and I settled onto a sofa that sat beneath the windows overlooking Troublesome and spread our lunch out on a circular coffee table. George Ella sat in an overstuffed chair across from us. It was a relief to take our masks off and visit indoors with her for the first time in more than two years. We could hear the rain pounding against the window behind us as we ate, but when we'd finished our lunch, the rain had let up, and we didn't have to use our umbrellas as we headed up the hill to the Gathering Place, where my class was scheduled to meet. I remember feeling relieved that I wouldn't have to lecture for an hour in soggy shoes.

At some point during my class, the rain started again, and afterward, as we left the building, I saw the porch festooned with bright-colored umbrellas. Reclaimed by their owners, they created splashes of color against the leaden sky as we all went our separate ways. Sam, George Ella, and I paused on the covered porch in front of the Settlement School office to say our good-byes. The office space had formerly held what I'd long referred to as ship berths, a cluster of built-in bunks that I'd avoided for decades because just the sight of them made me claustrophobic. When I glanced through the office's new glass double doors, I saw that the berths were gone. Desks and computers occupied the space that in years past had been the site of many informal late-night readings. The improvements to the ground floor had taken a great deal of money and effort, and I felt sure they'd serve the settlement well for decades to come.

Having hugged George Ella good-bye, Sam and I walked to our car. It was parked under what I still called the new bridge, a concrete structure that sat high above the creek to carry traffic in and out of the campus. Although not as new as the other campus renovations, the bridge hadn't been there during the workshop's earliest years. At that time, when you turned off the main road, you drove down a steep bank, over a narrow bridge that sat just a few feet above the creek, and up the opposite bank to reach the campus buildings. That entrance had been sufficient for more than eighty years, and I'd mentioned all this to Sam as we drove over the new bridge.

By the time I'd finished reminiscing, we'd reached the parkway and were headed home to Knoxville. The rain was coming down harder, and we had to turn the windshield wipers on high. Thirty-six hours later, when I switched on my computer at home, I watched in horror as my Facebook feed filled up with photos of devastation. I read post after post from writer friends who spoke of escaping to higher ground, of watching their cars being swept downstream from the spot where Sam and I had parked just days before, and of the pervasive smell of gasoline amid the floodwaters.

Troublesome had crested at a height at least ten feet above the previous high-water mark. Water had flooded into the apartment where we'd so recently eaten lunch. George Ella and her roommates had fled in the middle of the night as water poured into the hallway and swirled around them. The water reached chest high in the office space, surging with such force across the porch where we'd sheltered from the rain that it burst through the office doors. Floodwater had eventually filled the entire ground floor, leaving wreckage and ruin.

While I continue to rejoice that the workshop participants and Settlement School staff all survived that horrific night, I also deeply mourn the loss of life and the upending of livelihoods that devastated so many towns throughout eastern Kentucky. As for the Settlement School, it was as if all those wonderful renovations had never occurred.

This flood was an Appalachian apocalypse that had been decades in the making. When I look at the photos of the damage visited on eastern Kentucky in general, and the Settlement School in particular, I hear in my mind words of warning spoken in the Great Hall sixteen years earlier. "God made these mountains," one resident had said, pounding his fist on the table. "Y'all have no idea what will happen to this place once those mountains are blown to bits."

Neema Avashia

Fight from Away

"There are no jobs here."

This lesson came from my father, who, most nights at dinner, would direct the conversation toward employment or lemonade-stand economics. Jobs, after all, were the ostensible reason he'd left India in the first place: for the income and opportunity that work in the American labor economy provided. When he moved us to the Kanawha Valley in the early 1970s, there were jobs aplenty at the chemical plants that dotted the banks of the Kanawha River and at all the businesses associated with that industry. Jobs that could move working-class Appalachians and working-class immigrants alike into a growing middle class. But by the mid-1990s, when I hit middle school and started to get these lectures, the employment picture had shifted. Plants shuttered. Jobs relocated to other states or were eliminated entirely. And any kid of immigrants will tell you—when work disappears, our families disappear too. That's what happens when the only safety net you have is the one you're knotting under your feet as you walk the tightrope of financial security.

This lesson also came from my sister, seven years older than me, who went to an Ivy League college in upstate New York, a good thirteen hours from our home in Cross Lanes. When she came home during school breaks, she made lists with me of places she thought I might like for college.

None of them were in West Virginia.

When I pushed, even a little, and said things like, "Maybe I want to stay here" and "Maybe I'm happy just being average," she did not hide her dismay.

It came from my teachers, who told me two things: "Don't become a teacher." And "Don't stay in West Virginia."

And it came from the Appalachian atmosphere of my childhood, replete with both the smell of chemicals and the notion that you need to leave to succeed.[1]

I didn't interrogate those messages, because no counternarrative was offered to me as a young person and I hadn't yet figured out how to build my own. "Stay and fight" wasn't a message I'd heard anywhere. "Go back to where you came from" was a message I'd heard all too often. People who loved me were telling me to go. People who espoused xenophobia were saying the same. Consequently, I internalized the need to leave as a requisite for both success and survival. I left for college in Pittsburgh and only continued to move farther away afterward: to Madison, Wisconsin, for graduate school and then to Boston to work as a teacher. But even after two decades in this frigid New England city, I don't know how to call it home.

Appalachia is home. Appalachia will always be home.

Growing up, I lived on Pamela Circle, a street with neighbors who constantly sought ways to support one another. The kinship economy, Ann Pancake calls it—this recognition that no one is coming to save us, that we must depend on one another to survive and thrive. That looked like my dad giving physicals to every kid on the street before basketball season, and Mr. Starcher coming over to help with oil changes, and Ms. Carney and Mom swapping Christmas-cookie plates. It looked like needs being met without anyone even having to utter the needs aloud.

Since I moved away, it's been incredibly difficult to find a community that models a similar ethic. Maybe it's city living, or maybe it's a particularity of puritanical New England culture, but people up here don't make themselves vulnerable to one another easily. They don't reveal their need. To see it, you have to look and listen closely, and that's become a hallmark of how I live my life here. Looking and listening for need and thinking about how I can meet it. As to whether my own unexpressed needs get met by those around me? Well, that's a different question.

In finding steady employment that paid a good wage, I had satisfied my father's metric of success. But when it came to my own metrics of rootedness and relationship, I did not feel successful at all.

Many of my relationships from childhood are far weaker now than they were when I was growing up. Folks have moved away from the street and moved on in their lives. Truth be told, in the years leading up to the publication of *Another Appalachia*, I'd started to feel like home was more a construct of my mind than an actual place I could go. When bad things happened in Appalachia—floods, mine collapses, chemical spills—I would watch and grieve from afar, but aside from donating to the GoFundMe of the moment, I struggled to know how to engage.

In the summer of 2019, I went to the Hindman Settlement School to attend the Appalachian Writers' Workshop. I had a lot of trepidation before I went. Could I claim an identity as an Appalachian writer? Even if I did, would people see me as such? It took me a few days to get it out of my head. But in eastern Kentucky, on the banks of Troublesome Creek, I found home again. In people who sounded like home, and extended care like home, and wrote stories and poems and essays filled with ways of knowing and being that were just so deeply familiar. In people who modeled that radical politics and rural life can go hand in hand. That you can, indeed, stay and fight. That rootedness is not stagnation; it is strength.

Hindman quickly became its own kind of Pamela Circle, with Appalachian writers offering me a literary home in adulthood that made up for the childhood home I'd lost to out-migration and decline. Folks I met there came to my readings, interviewed me for journals, served as conversation partners, and looked for ways to help me on my publication journey in the same way that the Carneys and Starchers and Withrows had helped our family growing up: see the need and meet it. For the first time since leaving West Virginia, I felt like I had new relationships, new connections binding me to Appalachia. Like home wasn't just a construct in my mind or a manifestation of nostalgia but, rather, a living, breathing place.

In the summer of 2022, I traveled back to Hindman to teach at Ironwood, a new program that focused on creating space for young Appalachian writers to build community and refine their craft. Together, we wrote essays, poems, and stories on the bridge over Troublesome Creek and in the chairs on the porch of the Gathering Place. Sarah Kate gave us a dulcimer lesson in the Great Hall one day, we disco square danced in there on another, and got to see enormous raptors on a third. We made the requisite pilgrimage to Yoders, with all of us piling into a few cars to stock up on molasses cookies, Ale-8-One, and Grippo's chips. We ate tomatoes from the greenhouse and bought treats from the Knott County Farmers' Market. And outside these explorations of place, we participated in workshops with incredible writers like Marianne Worthington, Robert Gipe, and Frank X Walker, all messaging to young people that their stories mattered, that their lives had worth, and that they were held and loved within this community.

It was difficult to leave at the end of the week. All of us, young people and adults alike, knew that something within us had shifted because of our time by Troublesome.

And then, just a few short weeks later, I watched that site of intergenerational healing and care be ravaged by floodwaters. I followed posts on social media from friends describing rising waters, washed-away cars, and their struggles to find a way to higher ground and then to their homes across Appalachia. I pored over photographs of the aftermath: the drenched archives, the ruined library of Appalachian literature, the luthier's workshop and homes and stores and libraries and schools laid waste by uncontrolled water.

But I watched from away. Because that's where I live now. Away. And watching the floods was like losing Pamela Circle all over again. Like home slipping through my fingers, me unable to hold it tight. But unlike Pamela Circle, where I've lost most of my direct connections to the community, in this case, I watched eastern Kentucky come out for Hindman and Hindman come out for eastern Kentucky, with offers of food and water, shelter and support for cleanup, archive restoration, and rebuilding. And it became my job to think about how I could offer support from away.

Among the many lessons I learned from studying the aftermath of Hurricane Katrina, one in particular felt especially relevant in this moment: Do not be the asshole who spends money on plane tickets and rental cars and strains local infrastructure because you have some emotional need to show up. Take that money and put it in the hands of the people already on the ground, who have the relationships, who understand the immediate needs, and who have been doing the work all along. So my first step was to reach out to folks like Gipe and Kelsey Cloonan, who told me where to direct my expatalachian efforts.[2]

I do not have rootedness; this is true. But what I do have is a very, very loud mouth. And some recently published material to use as collateral. In the days after the flood, I used whatever platform I've built to amplify the work that was happening in and around Hindman. To offer up copies of my book in exchange for funds donated to Hindman and Appalshop and eKy Mutual Aid. To pull the eyes of people in my Boston day-to-day toward the mountains of Kentucky and toward the people who live there and to not let them look away. To curse out any fool who said something like, "This is what happens when you vote for Mitch McConnell." It wasn't enough, but it was extending care in the only ways I knew how. And it's a template I've continued to work from since the floods: when need arises in Appalachia, I turn up the volume on my love and use copies of my book to generate attention, offer solace, or raise funds.

No one taught me what it looks like to fight from away; I was taught to not look back when you leave a place. This work of walking in Boston with my head constantly turned toward Appalachia is work I learn to do anew each day.

Every time I do a reading from my first book, a collection of essays describing what it was like to grow up at the intersection of queer, Desi, and Appalachian, I get the same question. Someone in the audience asks a variation of "Do you plan on moving back to Appalachia?"

And every time, I have to gather myself before I answer. When I'm in Appalachia, sometimes I am holding back tears when I answer.

Because the answer is no. No, I can't move back to the mountains where I feel like I have access to a whole part of my body that's shut off from me when I am in Boston. No, I can't move into a space where I will be fighting alongside radical Appalachian organizers who think the way I do, talk the way I do, and fight the way I do. No, I can't move back to the only place I've ever felt a sense of home.

I can't because I look at my partner and my child, and all I want, with the entirety of my body, is to keep them safe. And I look at the laws being rammed through legislatures in West Virginia and Kentucky and Tennessee—laws that seek to render us nonexistent. Laws that would deny my child the right to read freely, to talk about her family, to explore her identity fully. Laws that would punish my partner and me, both educators, for our inclusive classrooms. Laws in future sessions that may seek to nullify our marriage. To take our child from us. And I cannot figure out how to move home. I know so many queer folks in Appalachia who are choosing to stay and fight, but I do not have the strength within me to be one of them. If rootedness is the marker of being Appalachian, then perhaps the marker of being a child of immigrants is that I am always on the lookout for the greenest pastures, and once I find them, I'm not likely to leave.

Leaving hasn't yielded success, but it has yielded a modicum of safety.

I in turn asked a similar question at a reading this fall at the University of Charleston, just fifteen miles down the road from where I grew up. I stood in front of a room of twenty-five college freshmen, most of whom were from Appalachia, and asked them to raise their hands if they intended to stay in Appalachia after graduation.

Two raised their hands.

I then asked the students to talk about their reasons for leaving. Many gave answers that were predictable: They sought to move to places with more diverse demographics and social opportunities for young people. They had been told, much like I was told, that they needed to move to seek quality employment. They struggled with the political climate of West Virginia, with a legislature that did nothing to meet their needs.

One young woman's response has stayed with me. "My family has lived in Appalachia for eight generations," she said. "No one leaves. It's just embarrassing. I don't want to be stuck the way they are."

Suddenly, I was in the very moment that I'd wondered about for myself so many times. The moment when someone could offer a different lesson from "leave to succeed." But what kind of hypocrite would I be if I told her to stay while I chose the comfort of my Boston apartment? And would she even hear the "stay and fight" message coming from me?

I rejected the leave-stay binary that had defined my relationship to Appalachia for so long. Launched into a mini-lecture. "Look, the first thing I want to say to you is that staying in one place doesn't have to be a source of shame. It can be a source of power and pride to be so bound to land in a country where place is increasingly being erased of its character and where people are being increasingly stripped of their land. Rootedness feels to me like the deepest form of resistance. So if you decide you want to stay, be proud of the decision to do so. But I also understand the pull to leave. If you leave, then I want you to know that leaving doesn't have to be forever. You can leave and come back. You can do that ten times over. No decision has to be permanent. And last of all, if you decide to leave for good, then I'm going to ask you to think really hard about what it means to fight from away. Because this place, and the people who live here, deserve our support and allyship no matter where we live."

It was a lecture. And as a teacher, I know that lecture doesn't work. That the likelihood of those words landing in that moment was slim. But maybe I was saying them as much to myself as I was to her or to any of the young people in that room. Maybe the words I'm trying to formulate, always, are the lessons I learned growing up in Appalachia: that the creek will keep rising, that no one is coming to save us but us, and that our survival is going to require us to learn to fight in new ways. That leaving is not the marker of success; fighting is.

Notes

1. Appalachian out-migration is not a new phenomenon. Leaving the mountains in search of work began in the early 1940s as the coal industry entered its slow and steady decline. In places like Charleston, West Virginia, the city closest to where I lived, the population peaked in 1960, when census data counted 85,796 inhabitants. In the sixty years since, the population has only declined, with just 48,864 residents counted in the 2020 census. The census shows that while the US population grew overall by 7.4 percent from 2010 to 2020, West Virginia's population dropped 3.2 percent in the same time frame, dropping in each age demographic except sixty-five- to eighty-four-year-olds.

2. The term Appalachian has sometimes been defined on lines so rigid that it makes it hard for folks who don't fall within those lines to fully embrace the term. Many different groups of Appalachian people who possess intersectional identities have created new terms for describing both their relationship to place and their relationship to self. For example, some LGBTQIA+ Appalachians refer to themselves as fabulachian, while some Black Appalachians have claimed the term Affrilachian to describe their identity. In the same vein, people who claim Appalachia as their place of origin but no longer live there may refer to themselves as expatalachian.

Melva Sue Priddy
The Shape Water Takes

It can come in a rush or in silence
with or without an alarm to roust.
Black or red or brown, no soil
perfectly safe. Saying no no no
won't make yes yes yes. Bless
the rain or curse the rain, there is no
human choice. Tie a canoe
to the chimney top or a raft to the porch
and you'll never be ready for the torrent
that comes. Car, truck, or van
at the bottom of the hill is no more safe
than you in the middle of the night
when floodwaters come silently or noisily
as that first slush is hushed
until water rushes high enough, raises
to sink pole beans, sunflowers, the porch steps,
then crosses the road to close off your only way out.
Your nightmares roust you, or your son calls,
perhaps the phone left on silent or in another
room, but you hear it if you've been praying
diligently through the drought. Torrential
rains three days in a row, the hillsides
saturated but the insurance lapsed,
you'll climb the stairs, the attic, the trees

higher behind the house. If the animals
can drown, so you, too. Mud slides,
roads cave, your sister washes away,
but you don't know that yet. Ten feet,
twenty-six feet, is there any difference.
Water rushes every window, every curve,
slowing for no one, builds as she
carries more weight, more weight
become debris. If you survive, you inherit
stories of all floods before. The thieves
who may steal afterward—
the least of your worries.

Wendell Berry
Making It Home

AN EXCERPT FROM *FIDELITY: FIVE STORIES*

The east brightened. The sun lit the edges of a few clouds on the horizon and then rose above them. He was walking full in its light. It had not shone on him long before he had to take off the overcoat, and he folded and rolled it neatly and stuffed it into his bag. By then he had come a long way up the road.

    Now that it was light, he could see the marks of the flood that had recently covered the valley floor. He could see drift logs and mats of cornstalks that the river had left on the low fields. In places where the river ran near the road, he could see the small clumps of leaves and grasses that the currents had affixed to the tree limbs. Out in one of the bottoms he saw two men with a team and wagon clearing the scattered debris from their fields. They had set fire to a large heap of drift logs, from which the pale smoke rose straight up. Above the level of the flood, the sun shone on the small, still-opening leaves of the water maples and on the short new grass of the hillside pastures.

    As he went along, Art began to be troubled about how to present himself to the ones at home. He had not shaved. Since before his long ride on the bus he had not bathed. He did not want to come in, after his three years' absence, like a man coming in from work, unshaven and with his clothes mussed and soiled. He must appear to them as what he had been since they saw him last, a soldier. And then he would be at the end of his soldiering. He did not know yet what he would be when he had ceased to be a soldier, but when he had thought so far his confusion left him.

He came to where the road crossed the mouth of a small tributary valley. Where the stream of that valley passed under the road, he went down the embankment, making his way, first through trees and then through a patch of dead horseweed stalks, to the creek. A little way upstream he came to a place of large flat rocks that had been swept clean by the creek and were now in the sun and dry. Opening the duffel bag, he carefully laid its contents out on the rocks. He took out his razor and brush and soap and a small mirror, and knelt beside the stream and soaped his face and shaved. The water was cold, but he had shaved with cold water before. When he had shaved, he took off his clothes and, standing in flowing water that instantly made his feet ache, he bathed, quaking and breathing between his teeth as he raised the cold water again and again in his cupped hands.

Standing on the rocks in the sun, he dried himself with the shirt he had been wearing. He put on his clean, too-large clothes, tied his tie, and combed his hair. And then warmth came to him. It came from inside himself and from the sun outside; he felt suddenly radiant in every vein and fiber of his body. He was clean and warm and rested and hungry. He was well.

He was in his own country now, and he did not see anything around him that he did not know.

"I have been a stranger and have seen strange things," he thought. "And now I am where it is not strange, and I am not a stranger."

He was sitting on the socks, resting after his bath. His bag, repacked, lay on the rock beside him, and he propped his elbow on it.

"I am not a stranger, but I am changed. Now I know a mighty power that can pass over the earth and make it strange. There are people, where I have been, that won't know their places when they get back to them. Them that live to get back won't be where they were when they left."

Maurice Manning
Planting Trees in God's Country

My people sot down, they say,
two hundred years before the beginning
of Time, being kicked out
of the place where they had been, never
to prosper there, only to toil.
And here, in deep woods and on hills
steep and rugged and rocky, they made
a hill farm, not to prosper,
but, in quiet hope, to survive,
to plant in the ground and feed themselves.
And that meant clearing a patch of land,
cutting the trees, breaking the dark,
original canopy in violence
to let the sunlight reach the ground.
All for a little corn and beans,
surrounded by the first beauty.
If this was done unthinkingly,
without a measure of regret,
I do not know. I have my thoughts.
We have to live with ignorance,
even, painfully, our own.
We also have to imagine the past
and believe we come from it, not
to undo it, but simply to imagine

and therefore belong, by opening
the ground. And then imagine shade
in summer coming to this place
again and birdsong in the branches
of heaven-reaching trees, living
ladders stuck in the ground to give
the future another rung of its past.
The invitation is to climb.
One thing I know about God's country—it's all
there is, and it's supposed to be alive.

Amelia Kirby
Collective Healing

IN THE DAYS AND MONTHS FOLLOWING THE EASTERN KENTUCKY flood of July 2022, I tried to help facilitate access to mental health services for people affected by the flood. Like all the recovery efforts, it was a patchwork response shaped by the lack of infrastructure for mental health and other systems necessary to sustain community and individual well-being. As we look toward emotional healing and recovery, we will have to reckon with the forces and systems that contributed to the catastrophe. To understand the scope of the trauma requires us to reckon with the traumas that preceded the floodwaters and the inadequacy of our response to those traumas. Hopefully, such an examination will allow us to consider our capacities as individuals and community members to build and sustain the social and psychological resistance that will allow us to look beyond mere survival.[1]

My great-great-grandmother died in Breathitt County, Kentucky, in the flood of 1939. My father recorded an oral history with his grandmother and aunt, and their recollection of the 1939 flood rings eerily true to the floods in eastern Kentucky in 2022. In 1939, a flash flood happened in the small hours of the night, and people woke to their houses filling with water. In my family's case, the water was ankle deep when they woke, and by the time they could get dressed and attempt to leave, it was waist deep; the house had shifted on its foundation and wedged the doors shut. One son was able to get out and was washed, bloodied, far downstream. By the time he could reach help, his mother and sister had died in their home.

Accounts from that flood tell of missing bodies and survivors stranded in treetops, of people so terrified they had to be pried from the fence posts they'd clung to for survival, of the loss of homeplaces and the reordering of how communities understood themselves. Then, as now, the extraction of resources from the region was central to the devastating rising water. The hills of eastern Kentucky had been timbered so severely that the torrential rains ran off the denuded ground and filled the valleys, drowning dozens of people in the night.

My partner, Robert, was at Hindman Settlement School the night of the flood in July 2022. I've been going to the Settlement School since I was six years old. My dad has been part of the old-time traditional music week for as long as I can remember. I know the physical place in my heart, and the school holds a similar significance for Robert. Six in the morning after the floods hit, I woke up to two texts from Robert, received at 2:00 a.m. and 2:30 a.m., respectively. "The creek is rising fast," followed by "Oh shit."

By the time I read his texts, the power and cell were long out, and it was hours before I knew if he was safe or how much devastation Hindman and the surrounding areas had experienced. I was glued to my phone, reading updates and seeing pictures I couldn't fathom. Floodwaters in Whitesburg swamped buildings where I'd spent years working. Houses floated down torrential creeks. The number of people reported lost rose.

To examine the mental health impacts of the floods of July 2022 is to look at the layers of historical traumas and harms that form the psychological landscape of eastern Kentucky. Doing so asks us to incorporate the long-term impacts of payment in company money in labor camps and the legacy of unnecessary deaths from breathing coal dust, a form of drowning in itself; it means examining the proven detriment to mental health that comes from a life spent in proximity to mountaintop removal sites and the uncertainty of having access to safe drinking water; the hollowing out of our public school systems; and the recruitment of death-making economies like prisons and landfills.

Six months after the flood, I sat in a conference room and listened as my colleague Mary Cromer clearly demonstrated the connections between the deaths in the flood and proximity to mountaintop removal. In her presentation, she showed images of her workplace at the

Appalachian Citizens' Law Center with floodwaters coursing through the first floor. Mary now works from temporary offices as she helps make the case that the harms of this flood can be directly linked to the extraction economy of Appalachia. We are forced to grapple with the deadly consequences of the industry that has given shape to so much of life in eastern Kentucky.

When we start from those traumas, what is recovery? What is healing? How far back do we go to look for the place to start accounting for the loss, for the harm?

Working with mental health professionals can certainly be healing. It can be helpful to find the language to describe to a therapist what it feels like to look at mud-destroyed family photographs for the last time, write a check to pay a mortgage on a home that doesn't exist anymore, pull a neighbor's body from a culvert, or hold one's crying child close every time it rains. But we live in a place where the mental health infrastructure—like the rest of our infrastructure—is threadbare where it exists.

Pathologies and labels will be applied to the lingering effects of what people have experienced. The official designation of posttraumatic stress disorder entered the lexicon of psychology in part because of the aftereffects of the deadly 1972 Buffalo Creek Flood in West Virginia, where 126 people died because of the collapse of a coal waste impoundment dam. In the national consciousness after that flood, the story that emerged about Appalachians was of an expendable people habituated to struggle and inured to pain. This story belied the work that the survivors were doing to grieve and to heal—forming support groups, committing to rebuilding even when told their homes were worthless, participating in lawsuits to hold the company accountable, rejecting the external valuation of their lives and homes, and building internal strategies for survival.

My friend Carter was also in Hindman that night in July 2022. In his book *The Evening Hour*, he describes a flood caused by a mine impoundment dam break. People scramble the hillsides to escape the rising tide of mine waste, and everything in its path is washed away.

The film made of that book was shot in Harlan County, Kentucky. The town of Harlan is surrounded by a floodwall, built after the flood of 1977. Outside that wall, across the river, there are now sports fields and tennis courts where there used to be the thriving Black community of Georgetown. The dominant story of Georgetown is that it was washed away in 1977. What isn't as often told is that developers were in the process of forcing people out and buying up the community when the flood hit. Rebuilding Georgetown was never in the cards. Its former residents are scattered throughout Harlan County and continue to tell stories of the richness, complexity, and abundance of that community. When we recover stories, we begin to recognize the legacies of loss, the relocation of traumas, and the possibilities of survival even through what is framed as utter destruction.

The film made of *The Evening Hour* features a song with lyrics describing celebrating destruction caused by floodwaters. I'd never paid those lyrics any mind until listening to them a few months after the flood. I sat back surprised, thinking how strange it was to write something so cruel. But on the heels of my first thought, I remembered my own reactions to seeing the flooding of places where I and others had experienced harm. There were moments in the days after the floods when I thought "Just deserts" and "You reap what you sow" and then felt guilty and strange. And if I'd had those difficult thoughts, what might others be experiencing? People who lost homes where they had been trapped in violent marriages or had been harmed as children? Sites of rape or hunger or the myriad of harms we can experience in our own homes. Is there a form of freeing in that loss? Does it release some harms even as it creates others?

In 2022, I sat with women who could precisely describe the last cries of their loved ones as the waves overtook them in the flood. Stood with a man as he explained the moment when he realized that the fabric he was looking at underwater was the last known clothing of a child now dead. Grieved with a friend reckoning with the loss of someone who had survived the flood but couldn't live through the aftermath.

What happened on Troublesome Creek, in Hemphill, on River Caney, in Millstone, in Lost Creek, in Ten Mile is a horror. And unfortunately, what has happened here isn't unique. Surely eastern Kentucky is its own place, with our own histories and our own particularities, but our fresh harms gather us closer to shared pain across the world. As climate change makes catastrophe a common occurrence, our story has joined the collective experience of communities across the world. If we are looking to survive the changes in our climate and our communities, we have to look to other places that have already been dealing with these changes.

When we think about our mental health—the capacity of our minds to help us live through the challenges we face—what stories do we carry in ourselves that make it possible to survive hardship? When we look for healing, we have to look to our collective community knowledge of how to keep ourselves safe. How we heal together. How we rebuild. We all too often think of mental health as an individual path, but I believe there is greater strength in building a collective understanding of how we help save each other.

Anyone who was in eastern Kentucky in the days and weeks after July 28, 2022, knows well how deeply collective the response to the flood was. People fed each other, gutted houses, hauled water and generators and bleach up the creeks and hollers, held story circles and listening

Moses Owens, maintenance foreman. © Tyler Barrett

sessions, and strengthened powerful mutual aid networks. The cumulative pain of this event will be with us for a long time, but so will the cumulative understanding of how those who survived found each other and built systems of support and survival that will carry us forward. How we helped each other carry the burdens of shared trauma and together built formations that will sustain our capacity for healing and growth.

When we're faced with the loss of the flood, we collect our power and build protections. We catch what's floating away and draw it back toward us. We strengthen the floodwalls and clear out the creek beds. We find higher ground and pull up whoever we can grab. We hold on tight. We look through the night to the dawn and the reaching hands of those we will rebuild with. We know it will never be the same. We make new stories of how we survived and how we love those who were lost. We follow the paths of those who have taught us loss and healing and how to hold devastation and anger and utter love. And we make ourselves ready for the next time.

Note

1. For further discussion of the concept of "social and psychological resistance," see David Denborough, *Collective Narrative Practice: Responding to Individuals, Groups and Communities Who Have Experienced Trauma* (Adelaide: Dulwich Centre, 2008).

Darnell Arnoult
This Too Is Creation

God's uncanny procedure in his hysterical solitude
is to night-bomb the mysterious void. Miles

of system-like threads pull the weighty direction of time. A rock
of light splits and shatters. What electric

silence floats on the eye of that mysterious radio. Rushing
waters cross mountains in human's crusty

sleep. Long-toothed mercy glows in the steel craters
of each face of nothing and something.

Water's Dark Body

Footbridge across Troublesome Creek with debris and a dumpster. © Tyler Barrett

Marianne Worthington

Rise and Fall: A Sonnet

> *the creek argues*
> *with the rain, grows bolder before losing*
> *itself, overcoming the banks that have*
> *defined it.*
>
> —Kathleen Wakefield, "Flood"

More than just an argument with the rain,
no, this creek was a broiling fracas. No
little spat, harsh words spit and regretted
later. This creek raised up its liquid fists
and knocked out sheds. This creek swallowed then spit
back out whole cars, campers, and guardrails. This
creek sank brown teeth into every crevice,
filled every cavity with its dirty
cascade, washed away keepsakes and photos,
sluiced families in two. This creek stood up,
enraged, a mad push of water higher
than we'd ever seen. Later, shrinking back
to its banks, a vestigial flicker
of itself—for now—reflecting the sun.

Kari Gunter-Seymour
Coal+iron+natural gas

a strange alchemy, hard to explain.
In Appalachia even rain comes at a price.
I stare out the window, nerves
unraveling, the river having jumped its banks
only inches from my door.

Clouds—silver-tongued, mouths
ripped open, spew swift currents.
Winds poke wily fingers
through the trees, squall obscure music,
a whine of breath over an open bottle.

The last of the potable water
fills a pitcher. Low light slithers
across linoleum like phantom
floodwaters, their shadows
a scatter of rats abandoning ship.

Palms cupped, I craft a shrine,
toss words I once found comforting
into the past like weeds, see myself
for who I am, who I've always been:

invisible, expendable, hillbilly, hick,
land-ravaged, beaten, left for dead—
wonder what's the longest I've ever
held my breath under water.

Tina Parker
How to Sleep

You'll wake in a sweat:
>*Who will die*
>*Who is dying*
>*Who is dead?*

Your jaw will be sore.
Massage each side.
Try an ice pack; some ibuprofen might help.

Most likely you're grinding your teeth.
You could be clenching your entire body
To prepare for what is next.

Carter Sickels
Troublesome Rising

Sometime around one in the morning on July 28, Matthew Parsons, poet and musician, arrived at the door in a rain jacket, cargo shorts, and Crocs. Drenched and wild-eyed, he looked like a fisherman who'd survived a storm out at sea. He told those of us still awake that Troublesome Creek was rising and that the cars parked under the bridge were in danger of washing away; his own was flooded up to the headlights, unreachable. Thunder echoed through the hills; lightning flashed and lit up the black sky.

I was with a group of writers hanging out at one of the cottages on the Hindman Settlement School campus, celebrating the fourth evening of the Appalachian Writers' Workshop, the premier literary gathering in Appalachia.

It had been raining for two days, and everyone in eastern Kentucky was aware of the storm. But the water rose so suddenly that our festivities quickly changed. That night, flash floods swept away entire homes and destroyed communities in Breathitt, Letcher, and Knott Counties. Since then, thirty-eight bodies have been found, and still more people are feared dead.

Floods in this corner of Appalachia are not new. But the frequency of flooding in the last couple of years, and the intensity of the storms, is unprecedented. That night, rainfall rates reached four inches per hour.

Troublesome Creek twists through three counties in eastern Kentucky and runs right through the small town of Hindman, the county seat of Knott County. It winds across the grounds of the Hindman

Settlement School, the first rural social settlement school established in America. Hindman is still active: it provides educational and service programs for the region and hosts the Appalachian Writers' Workshop, an annual gathering in its forty-fifth year.

On a typical spring day, the section of Troublesome that crosses campus is a picturesque Appalachian creek, no more than six inches to a foot deep and about an average of six feet wide. Troublesome and the bridge that runs over it are a touchstone for many who love Hindman and the writers' workshop.

Four days before the flash floods, I drove from Cincinnati to Hindman on a humid Sunday afternoon. For the week, I would be teaching at the writers' workshop, which draws emerging and established writers, most of whom have a connection to Appalachia or the South. For many, including me, traveling to Hindman feels like coming home.

When I turned onto Kentucky Route 80 East, I rolled down my windows to take in the verdant summer folds of mountains and wooded hillsides, threaded by creeks and streams. People here live in hollers and bottomlands, close to rivers and creeks, because the hillsides are too steep for homes. The Appalachian Mountains are some of the oldest in the world, rich with deciduous hardwoods, medicinal plants, and edible delicacies like morels and ramps. But the land has been decimated by decades of coal mining, specifically the destructive removal of mountaintops, which escalated in the early 2000s and still occurs. Mining companies rip away forests and trees and blow up mountaintops, removing up to eight hundred feet of rocks and soil from peaks to access coal seams. The earthmoving buries headwaters under rubble, poisons the ground and water, and strips hillsides of trees and vegetation, their natural defenses against flooding. Because of climate change, rainfall across the southeastern United States, including Kentucky, has increased by almost a third, and heavy downpours and flooding are more frequent. Strip-mine-damaged regions are especially vulnerable to hazardous flooding.

I've taught at the workshop since 2016, and I feel a profound connection to the place and the community of writers, musicians, and activists who attend. I grew up in central Ohio, but my grandparents and much of my extended family lived in Appalachian Ohio. My memories of that area, where I also spent my college years, run deep. Both of my novels are set in Appalachia. In my debut novel, *The Evening Hour*, a deadly flood, caused by a coal slurry dam breaking, hits a community in West Virginia. I read and watched everything I could about floods in Appalachia. But I never witnessed one. Until recently, I don't think I fully grasped the urgency of the phrases "higher ground" or "come hell or high water."

On the evening of the flash flood, there was a faculty reading, with about eighty attendants in the audience. Later in the night, a group of us headed over to one of the cottages on the hill to listen to music and talk, as we had every night. The rain was steady, and I'd noticed earlier that afternoon that Troublesome was higher than normal but still below its banks. I could have waded through it without much trouble.

I sat outside on the covered patio for a while, listening to the guitar, fiddle, and collective voices lighting up the dark. We drank bourbon and munched on Grippo's potato chips, talking and laughing and, as we say in Appalachia, having a big time. Through all of this, the rain came down.

What we couldn't see in the darkness was that Troublesome was swelling. After Matthew Parsons came to the door, a few of us tramped through the heavy rain to the main building on campus. The creek had spilled over the banks and into the lowest section of the parking lot, where two cars were about a third of the way underwater. My car was parked on the higher side of the parking lot, and I assumed it would be safe, but at the suggestion of the others, I moved it. Though Troublesome was rising, it was still difficult to imagine it climbing over the slope to flood the entire parking lot or the building where I was staying.

I went back to the cottage where we'd all been hanging out, and about a half hour later, we lost electricity. Emergency lights flickered on, and an explosion echoed down the hill—"A transmitter just blew," someone said. I looked out the front porch window and breathed in the distinct, unnerving stench of propane. All I could see was darkness.

My friends and I walked back down to the main building but could no longer get around to a spot where we'd been standing less than an hour before. The parking lot was now submerged. A white pickup bobbed in the swift-moving water, nose first. The alarm in the main building steadily shrieked. The lower apartments, where I'd had a room, had completely flooded, with sewage bubbling up from the drains and water rapidly rising from ankle high to waist deep. The three women who had also been staying in the apartments had made it out safely and had generously gathered up all my belongings.

I helped carry up musical instruments from the storage closet. About six or seven of us moved quickly, quietly, and carefully, guided by the flashlights on our phones and the dim emergency lights. Two Hindman staff members waded into the office, trying to save what they could, but the water was already too fast and strong—while they were there, the water burst through the doors and shattered windows, quickly rising to chest level.

There were about sixty people staying on the campus. People woke up those who were still sleeping, moved cars to higher ground, and filled up jugs and bottles with potable water. A woman I'd sat with at dinner asked me to shine a light over the lot, where her car was partly submerged. She wanted to try to get it out, but the lot had transformed into a sea of black water, with eddies swirling. I said, "You can't walk through there; you have to let it go." She burst into tears, and I held her. It was harrowing how quickly the situation had changed.

At one of the cottages up on the hill, where many of us had relocated, we gathered on the big front porch and stared anxiously into the black night. There were quiet conversations and occasional sparks of humor or sudden hugs.

Other than rumors and speculation, we had no idea how high Troublesome had risen or how far the flooding had spread. We had no cell service or internet. The rain pounded the hills, and we worried about mudslides speeding down the steep slopes. We wondered what would happen if the main bridge washed out. Many communities and homes in the mountains only have one road in and out—once the bridges go, it's impossible to leave, to find higher ground.

The night was a strange blur, both chaotic and calm. I didn't sleep. Time passed slowly, and we waited like children for morning to break through the darkness.

The light finally appeared from behind the ridge. The morning was disorienting, as we tried to grasp what we were seeing, how geography had shifted overnight. I followed a couple of others to a clearing to see the town. A picture on my phone taken the day before from the same spot captured Hindman's downtown, the quaint Main Street with its two- and four-story brick buildings, including the Appalachian Artisan Center and a music shop specializing in handmade banjos and dulcimers. Now, the shops were half-submerged, the green spaces and parking lots looked like lakes, and the cars had turned into boats that could not float.

Troublesome had turned into a mighty river overnight. The tops of tall trees still rooted in the ground rose up like gigantic lily pads, their trunks deep underwater. I watched the creek carry away a shed, an empty kayak, and gigantic branches. Busted boards, shattered pieces of buildings, and unrecognizable plastic wrapped around telephone poles and the iron bridge on campus. We didn't know if the roads were drivable. We waited and tried our phones. A few thoughtful people scrounged around in the kitchen and set out food, including a leftover Butterfinger cake and biscuits. The alarm that had been going off all night continued to beep incessantly, but, exhausted and heartbroken, we ignored it. Down by the main building, a strange flutter of white came into focus: four big ducks waddled back and forth, squawking and distressed.

Around nine that morning, the rain slowed, then stopped. We heard that the road through town had opened one lane. We packed up our cars, and rides were arranged for the ten or twelve people who'd lost their cars to the flood. We quickly, sadly said our good-byes. Most of us live elsewhere: Lexington, Asheville, Brooklyn. It felt hard to drive away—from my students, the staff, and the people who lived there, facing such immense loss.

Back at my home in Cincinnati, I learned, along with the rest of the country, the scale of the catastrophe. Four days after the flood, thirty-seven bodies had been recovered, including four young children from the same family. Governor Andy Beshear announced that others were still missing. All over the mountains, terrified people perched on rooftops and waited to be rescued by helicopters. Creeks, streams, and the North Fork of the Kentucky River had risen up to twenty-five feet in some places and ripped a path across eastern Kentucky, swallowing homes and devastating little towns. Over in Whitesburg, Appalshop, an arts and culture center and hub for filmmakers and musicians, stood under six feet of water, the archives soggy and mud drenched. Six buildings at the Hindman Settlement School were severely flooded and damaged, including historical buildings and the office that housed archives, precious letters, photographs, and works by authors and musicians. They lost computers and desks from the learning center, and all the produce in the greenhouses was destroyed. But the footbridge over Troublesome remained intact.

Ten days after the flood, I went back to Hindman to bring supplies and donations I'd collected. It was raining yet again. In town, piles of debris sat outside ruined shops, with the stench of mildew thick in the air. The power was still out. But the school, like other community centers across the region, had become an emergency outpost, providing hot meals, water, and supplies. The rooms where we'd had readings and lectures were filled with tables of nonperishable food, cleaning supplies, dog food, toiletries, extension cords, and batteries, all free for the community. I saw staff and volunteers, including FEMA employees, busily working. Pallets of bottled water sat in front of the main building. Two older women pulled up in a pickup and loaded one case of bottled water in the back. The women asked how many cases they could have. "As many as you need," the volunteers told them. They took only two more.

Appalachia has suffered from years of neglect by national and state politicians, with decaying infrastructure, few job opportunities, and scarce educational resources and health care. Decades of coal mining have left communities exposed, without natural defenses against future storms and floods. Nearly two weeks after the floods—after voting against legislation that would invest billions to fight climate change—Senate Minority Leader Mitch McConnell and Senator Rand Paul, both representing Kentucky, finally visited the region.

I am writing this as a witness to climate change and corporate greed but also to the resilient people all across eastern Kentucky who are helping their neighbors. Community members, activists, and volunteers are working long, exhausting days to salvage what was lost, to provide food, water, and shelter. I feel buoyed by the love and care I see in the community. I want to say that this can't happen again, but a declaration of that sort should have been issued years ago, by those with the political power to alter years of ecological neglect and economic greed. The flood came at night, forcefully and quickly, destroying so many lives in its wake. Unfortunately, I'm afraid it will happen again and again.

Jesse Graves

From the Tennessee Side of the Mountains

I watched the storms moving without knowing
what I saw, and even the lightning looked wrong
from the Tennessee side of the mountains.
I was dry, under the shelter of Scott's carport
with Josh and Kevin, talking into the later hours
and sipping the terroir of Kentucky straight.
Scott said, "I don't think that's heat lightning,"
and I said, "It reminds me of New Orleans,
how storms would march across Pontchartrain."
We wouldn't learn for hours what we had witnessed,
and the next days were filled with an anxious dread,
as numbers of drowned and missing kept climbing.
I listened to Bruce Springsteen's *Nebraska* to hear
how the band played "Night of the Johnstown Flood,"
casualties still rising as waters began to recede.
I whispered helpless prayers for Troublesome Creek,
for the small children swept from their parents' arms,
and for those whose stories we now will never hear.

Jamey Temple
Fire and Rain

In the course of a tornado warning, a house is struck by lightning, catches fire. The same week the following year, midway through a writing workshop, heavy rains fall, cause a "once in a thousand years" flood.

I.

It begins in sleep, when safely tucked into a bed that holds you while you dream.

It begins deep in the night. When the skies can't be seen, where the only light is from licks of lightning.

It begins with sirened beeps, *danger*. Get up, move. Move lower, move higher, always move.

II.

During, you wait. In a basement, your eyes dart from corner to corner, the house shaking from the storm raking over her. When there is a pause, you wonder if the worst is yet to come, if the swirl of debris will find you. Then a boom. Sparks dance across the metal ducts that outline the room.

Before you emerge, you have a moment when your body relaxes. You're okay. No swirl has swept you up, just a strike of electricity from the sky.

You smell smoke.

Before you think about what to grab, you run up the basement stairs, out the door, and down an exit ramp, away. You don't want to turn back, to see the flames unfurl, char, and melt the place you call home. The fire truck's siren screams louder; its red lights stroke the black.

During, you wait on higher ground. You don't think about everything you left behind in a place that mud could overtake and make slide into the floodwaters below. You don't want to see the high water, the lightning's flash ripple on the current.

Others watch from the porch, smoke, huddle. You stay in the womb of the house, fetal, telling yourself you're okay. You don't want to know how close that water is to you or what it has taken as its own.

The thunder rumbles like a dining chair scooted across the floor. The room smells damp, the moisture sticky and thick. A dog yelps from her temporary cage of a bathroom. It's the only living thing that can cry.

III.

After, you stand at the fence to retell the story, to say all that matters is that the people in your house are okay, to answer questions about what you need. You think you need nothing. Isn't being alive enough? You kick your foot against a post to which an ant clings. You should be still talking, thanking others for caring, but now you can only think about ants. How as a child, you studied the mounds of dirt, examined each soil pebble that built small pyramids, the worker ants entering a center hole. They marched in line, orderly, safe, until you pressed your flip-flop on top, flattened their progress. They scurried, with nowhere to hide, nowhere to go. You'll understand much later that below, their networks were intact. They busied themselves to recover, but only after their distressed cries.

Their mounds popped up again, right? They always came back.

After, at daybreak, you walk out of the house ready to finally see. The waters still high, a family of ducks flits here and there on the banks above. Cars pocket where the waters have receded, cars with trunks popped open, flashers flashing, windshield wipers flapping.

But what you'll remember is the crayfish displaced on the bridge, caught in the floodwater's puddle.

Standing on the bridge over Troublesome Creek, you don't see the crayfish yet. Below, debris bobs out of brown water the color of paper grocery bags: swimming pool noodles, a car flipped and stuck, specks of red, yellow, blue. You turn to walk away from the muddle, head back to your car, which you'll soon pack to leave, when you finally see him, small and nearly camouflaged.

The crayfish's pincers move. You don't know how to help him: If you pitch him over, he'll surely die. If you leave him exposed, he'll die too.

IV.

It never ends, renews with each dark cloud and drop of rain, the smell of smoke that still breathes from your house in summer heat. You'll always jump when you hear a siren or storm warning; you'll always look for exits, even when you're not ready to leave.

You've seen fire. Rain. Swollen creeks. People you don't know carry out your smoked, soaked belongings. People from here, there, everywhere who load their vehicles and come to you, to others, as if there were a distress signal they were meant to answer, to carry you when you couldn't carry yourself.

Jayne Moore Waldrop
Before

Before we arrived in Hindman
the forecast hinted at rain midweek.
I slipped a tiny purple collapsible
umbrella into my book bag thinking
that an umbrella remembered insures
against clouds. Little did we know
what was coming. Always before

we'd known the creek could grow
troublesome when too much rain
pushes it out of its banks
to spread across the low field,
the bottomland, before settling
back down to its rocky bed.
Little did we know that the water
would rise but also come in all
directions, nonstop, sideways
on fierce wind, down the mountain,
carving new rivers from earth
alongside the cabin where we
bunked, before a trickle came
through the ceiling drywall,
a rivulet through light fixture
and A/C vent, soaking bed linens,

TROUBLESOME RISING

drumming into trash cans,
finding its way, before

emergency alerts surged on phones,
lit up our rooms with warnings
of life-threatening conditions
all around us, warnings to stay put
if currently sheltered, warnings
that the rain wouldn't stop
until morning, warnings as knocks
on the door, dire announcements
that the parking lot was underwater,
cars and motorcycles washed away.
We threw on clothes, cinched meds
and devices into plastic bags in case
we had to make a run for it, in case
we had to cross raging water, in case
the bridge washed out, in case
the mountain slid down on us, before

I realized I was not prepared for what
was coming. I had come to campus
in sandals and linen on an unclouded
hot July day. I had packed no rain gear,
not even a jacket. No boots, no rope,
no flotation device, no boat, no paddles,
just a tiny purple collapsible umbrella
that made sense in normal times
but seemed downright ridiculous
as armor against a storm overtaking,
cutting its own paths in ways
and places not troubled before,

before creekside rooms filled and folks
evacuated, before the lights went out
and we sat in the dark, before water
washed over history and people, before

Jayne Moore Waldrop

we witnessed sheds and oil tanks
bobbing downstream, before we learned
that floodwaters smell like muddy,
petroleum-based sewage, before towns
drowned, before we saw sets of steps
carried far from what had been home,
before extremes became commonplace,
before floodplains had to be redrawn,
before tornado alley shifted toward us
and set its sights on Mayfield, Bremen,
and Dawson Springs at the opposite end
of the state, before hundred-year
catastrophes came annually, before
drought shut down the mighty
Mississippi, before each year grew
hotter than the last, before oceans
warmed, before ice sheets melted,
before we admitted we humans
caused the changes, before we realized
we're not prepared for what's coming.

Patsy Kisner
Flood-Watching Instructions

Lean onto the porch railing,
find a grip, steady
yourself as you watch
the breeding waters feed.

Lift a light so you can better
see. Startle when a deer rises
from the swirling edge, wet
and slick, slow from stupor,
yet her large, sorrel eyes blink.

You gasp at first, that rush
of relief, but then you swallow
air, shiver harder, cry
because it's summer

and a doe
shouldn't be
alone.

Meredith McCarroll
Belonging

As dawn broke after the longest night and the waters started receding, I stepped outside. The landscape was all wrong. My mind tried to right the image. I rubbed my eyes. I tilted my head. The solid points of reference seemed all rearranged. What they were is underwater.

I saw Lyrae. We'd sung low and quiet earlier the night before, circled up in the shelter of a porch roof. I had recorded a video on my phone of our feet tapping in unison.

I walked toward her. Wordless, we embraced long and strong.

"My car swept downriver," I said.

She nodded.

We walked toward the water, headed downstream.

Few words. Taking in the jostled world as it resettled. Shaking our heads. "Mmm-mm."

As we crossed the bridge, we saw that the creek had risen to the bottom of the first floor, which stretched fifteen feet above the parking lot below. This morning, the creek filled in the space that had been a parking lot and barely flowed. The building itself had become a sieve, trapping trees, plywood scraps, and trash of all sorts. Movement caught my eye. A box turtle on a piece of wood in a pile of debris. A dozen feet above anything that would make sense.

Earlier, when we were moving cars at 2:00 a.m., we saw four ducks. The big white kind I only see in picture books. Not ever out in the world. They were in the parking lot, huddled together. Not sure where to go. By morning, they rested on the grassy hill below Stucky. Exhausted from escape, they slept close together.

Lyrae and I headed downriver. Behind the Mi-Dee Mart, we looked to see if my car was wedged this far down. At the back of the parking long, looking out at the flood, we saw an officer. Hat off. Eyes swollen and bloodshot.

He looked at us. Said nothing.

"Long night," I said.

"We've lost four," he said. We stepped closer.

"Two of our neighbors. They stood in their doorway, and they were calling for me. And then they were just gone. It went." He gazed somewhere beyond us. "I can't find them." He looked at us hard.

Lyrae stepped forward with her arms open. He stepped in. She wrapped her arms around him. He lowered his head and gave two big sobs. Stepped back. Wiped his eyes with one hand. Thumb and finger meeting at the bridge of his nose. Forceful, purposeful.

"I didn't know them too good. They was four doors down."

We nodded.

"I'm so sorry," I said.

He nodded.

We walked on.

We could barely see Main Street. I'd gone on a few runs there during the week. The arts center windows had blown out. I wondered about the luthier's workshop across the street. There was about six feet of water there now.

We turned back. Walking upstream, I counted four mudslides. An outbuilding smashed near the lot where we'd sat Sunday watching Carter's film. That first night, we had eaten popcorn—freshly popped by the new popper they were eager to put to use. Before the screening, a truck with shaved ice helped turn our tongues blue and green and bright red. Deep in the movie, during an outdoor scene, I realized that the sounds of crickets were not in the film. They were all around us as we watched. I realized I don't hear crickets where I live. I took a deep breath and decided to remember that moment. With salt on my lips and my jeans damp from sitting in the grass, I listened to crickets in the company of writers.

Now all of that place we'd sat to watch, which hadn't even been in sight of the creek, was the creek. A dozen feet deep and filled with gasoline and cars and animals seeking dry ground to rest.

My phone had died sometime around dawn. I realized that my partner might be worried. I borrowed a phone and left him a message. He called back, and someone passed the phone to me. When I heard his "Mere?" I broke. "I'm OK," I said, and I finally let myself cry. I tried to describe what had happened. And despite his warm response, I knew then that he would never understand. I knew then that I would get on a plane and end up in Maine and carry this quietly within.

Another writer offered to drive me as far as Knoxville. Too stunned to sleep and too exhausted to talk, I barely remember the drive. It may have been two hours. It could have been six. I arrived at the Hyatt downtown, where the receptionist said, "We've got a Starbucks in the lobby. Our rooftop bar opens in an hour." I nodded, signed my name, and took my key.

It wasn't when I thought of the snapping turtle eggs they'd pointed out, which they were protecting with traffic cones, now washed away, that I cried.

It wasn't when I saw the rental car floating downstream that I cried.

It wasn't when Josh and Sarah Kate returned from the office where they realized all was lost and the doors blew off and the glass shattered that I cried.

It wasn't when one writer cried because it was the anniversary of her house fire caused by lightning that I cried.

I cried when I opened a door into a fresh clean room with a big bed with the whitest sheets. It was then that I dropped my bags, fell to the floor, and cried.

Not for me.

I cried for everyone not driving away to check into a hotel.

I cried for the men picking through the debris on the road.

I cried for the officer who had watched a neighbor wash away.

I cried for the gas leak in the water.

I cried for the first editions washing away and waterlogged.

I cried for Tamela, whose motorcycle was gone.

I cried for the janitor I'd seen polishing the elementary school floors at 7:00 a.m. on Wednesday.

I cried for the women looking for their neighbors on the bridge.

I cried for Appalshop, also underwater.

I cried for the banjos and mandolins washed away from the luthier's workshop.

I cried for Mandi, who said, "This happens all the time."

I cried for eastern Kentucky, where floods and spills and collapses and catastrophes happen and no one cares or understands that mountaintop removal and climate change are the causes.

I cried for thousands of people I don't know but love, who have lost family photos and wedding dresses and quilts and kittens and gardens and maybe hope.

I picked up my phone, needing to tell someone I was OK. I didn't know whom to call who would understand. I needed to share this with those who'd been with me. Whose muddy shoes had their own tale. Whose soggy jeans were hanging in Boone or were shoved in a bag en route to Brooklyn. Whose bags of books were in another hotel room, waiting for roads to clear, for creeks to lower. I set the phone back on the bedside table and felt like an outsider in both worlds I inhabit. Back in Maine, Kentucky might as well be another planet. And Maine feels to me sometimes like a foreign country. But I also am not from Hazard. I don't know floods. I have never lost like that. When I got in a car to get on a plane to go far away to a place where no one is even thinking about a flood and has never heard of mountaintop removal, I wondered, Where do I belong?

Yet . . . I witnessed belonging as it was forged. When Lyrae, a Black woman raised in Florida and teaching at Cornell, reached out to the white officer, belonging was made. When I, a white woman from North Carolina who was teaching in Maine, spent a few days in the Hindman community, belonging was made. Forged by experience. Forged by values. Forged by intention.

bell hooks writes of grief and loss yet turns to hope in this poem from *Appalachian Elegy*:

> knowing that we
> have made a place
> that can sustain us
> a place of certainty
> and sanctuary

Hindman is sanctuary—a place of certainty. Appalachia is a place that can sustain us.

When, months later, I reconnected with writers from that week, all the images I'd swallowed down came rushing back up. That sob I'd held back until my chest burned came screaming out of my mouth. I felt I belonged.

To belong is to be seen. To be heard. As I find my own sense of belonging in this place that isn't my home and in an event that was such an exception from my narrative, as I seek sanctuary in a place in recovery, I wonder how Appalachia itself can belong.

What will it take for Appalachia to be seen? For her stories to be heard?

Perhaps it is the work of holding us together—of providing sanctuary and offering belonging—that most defines this place. Yet, as mud slides and creeks rise and buildings collapse, at a higher rate than ever, how much can any place withstand? Offering sustenance. Offering belonging.

When I asked Lyrae how she did that, she said she is a Christian who is guided to "love thy neighbor" and strives every day to show forth that grace.

"But how did you know to step toward him like that, with arms open?" Lyrae said it was her teacher Lucille Clifton who taught her that. Lucille said, "Don't not touch someone who's hurting."

So Lyrae stepped forward.

In "cutting greens," Clifton writes:

and just for a minute
the greens roll black under the knife,
and the kitchen twists dark on its spine
and I taste in my natural appetite
the bond of live things everywhere.

The snapping turtles and the washed-away ripe tomatoes and the four children from one family. Four children. The bond of live things everywhere bringing us together to write and sing and stay up late in sanctuary, and to break and mourn and lack the language for belonging.

Lyrae Van Clief-Stefanon

Strangers: Flood Crossing

Are you all right? I walked with M to the edge of the water. Not sweeping but raging. Not washing but wrenching. A turtle clasps a clotted mess of grass, debris, tangled scraps, wisps churned together and bobbing against the bridge pillar. The rough current reeks of gasoline. M and I are standing behind the gas station, then suddenly, to the left, the officer is there too. When I ask, something changes in his face, but there are no words for how his face changes. He points, but he is pointing toward a gap in time. I walk over and touch him because I understand something about the gap. I did not know that water. Mud-milky and welting, lashing, the water struggles against the town; tears at it; tears it. The water ripped the doors off the building where I was supposed to have been sleeping but wasn't. Because my nephew had COVID-19, my room assignment had been changed. Up in Stucky on the hill, I was asleep through all the rising. All night, dead asleep. When I walked out of the bedroom, I wasn't sure I was awake. The living room was crowded with belongings, packed with people who had not been there when I went to bed. There was a little dog in front of me. I think I said, "Good morning," but no one spoke. I stood there like a ghost in bare feet and a white cotton nightgown, feeling strange even to myself. Wake up. I looked to my right out the window and saw the people gathered on the porch. There had been a party going on at the Gathering Place when I went to bed. Grippo's. G in the kitchen telling stories, hilarious. The music circle. B is singing "You'll Never Leave Harlan Alive," and it feels like everything in nature has stopped to listen to him. Utter stillness. The night loves his voice like I do. At the last verse, the winds arrive. Like a presence, they tiptoe in, then shift, audience, then participant. The winds sing. Then the rain comes down hard in sheets. And we are sitting in the rain in a gap the roof makes at the Gathering Place. Then I am in a deep rain sleep all night, dreaming the party has moved to the porch at Stucky.   In the gray light, the ducks on the hill look so white that they make the green lawn seem neon. Then I see Troublesome Creek. Overnight, a trickle has become not a river but a stranger.   When I touch the officer's arm, he is welling up. He points to where the gap moves, much more slowly than the current. I see but cannot see through to what he sees. There is no way to cross this water to get through to the place where he is pointing. There is a doorway over there: two people standing in that doorway. Do not try to cross that gap. I wrap my arms around him.

Robert Gipe
Wall of Water: A Story

WALL OF WATER, THEY SAID. COME IN THE DARK OF NIGHT. Raindrops big as quarters. Couldn't see from me to you. Like a tsunami, they said. One big wave that come in below us. All we got was rain. But I heard it. Cars bashing into rock. Trees snapping. Metal twisting into knots. People begging for help. Hanging out their windows. Crouched on their roofs. Said don't leave us. Said come get us. Said God's sake, don't leave us. Make you sick to your stomach. Sound of the world ending.

Water come off down there where the road jogs left. Where the grass is knocked flat. Where the blacktop's busted up. Fright Branch, they call it. Beanstalk Bottom. Water come off a mine pond hanging over our heads since they started stripping. Ponds above us on three sides. Ready to bust. Steal your dreams most ever' night. Yeah, you can see it. Follow that power line a quarter mile. Be a gate on your right. Hour to the top.

When you see it, you'll know. Two billion gallons o' water from washing the coal. Arsenic in it. Lord knows what all. Castle Rock up there now. Don't know the original permit holder. Old-timers could tell you. If you see them.

Know her real well. Said she's in the front room that night. Said the water come in a rush. Like a tidal wave, she said. Knocked the house off its foundation into a shed building where her old man works on cars. Knocked the shed into a trailer. Sent the trailer down the creek. Them blocks is where it sat. Trailer picked up a camper. Camper knocked a hole in the church. Water carried off the rest. House after house, all down the creek. Like dominoes, she said. Underpinnings and patios snapped off, hanging in the trees. People's vehicles, their way to work, their pride and joy, spinning like Hot Wheels in the water. I'll be surprised if they build back. Water scared and land scalped to the rock.

No, honey. Wadn't like that before. Water done that.

Was her grandson got her out. Waded waist-high water. Packed her above his head like a loaf of light bread. Thin man stayed in the trailer. Said he couldn't swim. Said he'd ride it out. You could hear him singing as he spun. "Ooh, That Smell" was the song he sung.

No, honey. I don't need nothing. All I lost was a couple dogs. Well. That boy would drink them pops. But them down the creek is what really needs it.

How could it not be the mines? A tidal wave right below a pond? Us sitting fifty yards up the creek getting nothing compared to that?

Ain't just us neither. She was supposed to get a check from that outfit giving away money, but the mail ain't run since that night. I went down to Hazard to pick up her check. Talked to a woman from off Squabble, feller from Thunderstick, couple in Breathitt County, old man on Third Creek. All say the same. Wall of water. Tsunami. Tidal wave. Ever'one living below a strip job. Don't make no sense if it ain't the mines. That's not how rainwater does.

No, honey. We wadn't surprised. This is about the least surprising thing in the world. Bitter harvest of a hundred years of a man getting rich giving no thought to tomorrow or the people who live there.

Yeah, you could talk to her. She's staying up Beegum. But she ain't supposed to talk to you. Not about this. Years back she sued Castle Rock. Got some money. Had to sign a paper saying she wouldn't say nothing about the company. Put a gag order on her. But I didn't sign no gag order. You didn't neither, I don't reckon.

Y'all wouldn't care to carry me to the dollar store, would you? I'm supposed to see a man about a dog. I'll show you a woman could use them diapers.

I don't know who her lawyer was. Hard to say what they was thinking. Maybe this time will be different. I heard the inspector wrote them up for the pond being too full, for too much silt, for not dredging it out. They're supposed to maintain them ponds. But you know how that goes. I told that boy this morning. This ain't a flood. This is a mine disaster.

But I appreciate you fellers giving me a ride. You want one of these pops? That boy can't drink 'em all.

101

Annie Woodford
One of the Sounds of Water

The creek rose up while we slept.
We woke to it rushing the edge of our camp,
no more spellbound moss, no more
    tumble of clear water or dragonflies stuck wing to wing.
Did the crawdads bury themselves
deep under stones? One slip,
and the yellow flood would take you,
    bruise you, knock your head against a rock.
White noise roared with the void, the sonic chill
of interstellar spaces, like static on the AM dial,
traces of suns dead for ages, collapsed to pinpricks
    heavy as the devil's tongue.

Melva Sue Priddy

From Furman

Upon waking at 2:00 a.m., other women, there are nine of us in this upstairs room, have scrambled to the two small windows facing the road. They clamber, crowd, murmur in awe, no distinct words. My bed is on the farthest wall. They stare out the windows, pull on shoes, as they tumble into shirts and start out of the room, alarms sound   out of sync everybody's phone   flood warning—imminent danger. As they move away, I walk to the window   look   but don't see   don't know what I'm looking for   then in lightning   farmers' market pavilion   across the road   water   up to the roof   up to the roof   impossible   white truck floats behind pavilion how possible   red car floats   how possible   I remember a woman in workshop the day before watching the rain. I asked if she was worried; she's local, said her daughter alone at home— her road had flooded out twice already this summer and it never had before. I remember asking if she wanted to go home early.   Pulling on shoes   I join other women   on front porch   thundering rain   voices muffled   waterfall rumbles   mountain side on right   heavy water curtain   falls running   straight down mountain   floods yard 4 inches roadway on left   holler behind us   releases mud water   churns asphalt   but I'm not sure   ask   is that buckled asphalt   the three sharp upsurges   nearly height of car   yes   yes   within half hour electricity out   flashlights   there used to be flashlights in all rooms but none   cell towers may go   young women scramble   to save fresh water inside   bring trash cans out   catch rain water   I think out loud flashlight in car   boots in car   young woman says I have boots on retrieves boots from back seat   I scrounge   four flashlights   stashed in trunk   bless husband   split between upstairs   downstairs   two on porch   to watch the water level   but none of us go back to bed   cars already moved   upward lot   someone says   we/furman   high enough   but we plan to move up   mountainside   if water starts to cross that road   I call husband 4:00 a.m.   reassure him I'm ok   he often stays up late watching TV   don't want him worried   waters flood first-floor library   across the road   we're cut off from   settlement school's main buildings   cut off from main body of people   but musician   who woke us   after night party ended   drives back and forth   tells us   some cars flooded   main street hindman   flooded   then whole of hindman flooded   no open routes   bridge never flooded   flooded   houses collapsed   homes washed away   roads washed out   vehicles washed   a bad odor   none of us recognize   gas   diesel   oil or what   we don't know   after daylight   we walk across road   survey damage   Someone calls us   to opened-refrigerator breakfast   Are you ok?   Are you ok?   Are you ok?   hugs   hugs   hugs   hugs   tears   We repeat our own versions of what we have seen   repeat versions of what we remember   repeat our worries.   Cars had been pulled   up the mountain road to Preece   bumper to bumper.   Ground floor of Mullins Center destroyed.   We wait for floodwater to recede   road closure notices.   Most EMTs and safety officers flooded out,   three people known dead.   talk   talk   talk   Tension in laughter.   Wait. Wait.   Wait.   Released   drive out earlier than expected.   We secure rides   for those without transportation   only one safe route out   slowly drive past devastating scenes   main street all windows broken   water halfway up storefronts   only one lane dug out across   one bridge out of hindman   opposite side of road   whole neighborhood flooded   never seen water this powerful   I call husband   turn left on 80   will not try   jackson route 15   body suddenly jittery   stop to breathe   before continuing homeward   I won't make it   call friend in berea   I don't even know where I am   Come, she says   stop here to rest   and I do   I think I'm in shock   talk   talk   talk   soup   sleep for an hour   so many people have lost so much   this I know without   being told   going home   when so many have lost homes   every   thing   then tune in to news images   beyond that morning's reach   drifting images   brown watershed images   sunshine   green

grass   green trees   destruction   deaths mount   deaths mount
descriptions mount   can't sleep   vow no more news   but people lost
children   homes   lost so much   mudslides in bones for days   others lost   so much loss   beyond understanding   the shock   in
my body   for days   unable to talk without   hurting   and I don't
even live there . . .

Displaced ducks from Frogtown. © Tyler Barrett

Tia Jensen
Turtletalk

THAT FIRST NIGHT, YOU'LL FIND ME CURLED UP IN BED, UNDER AN old comforter, tucked into the top-of-the-stairs room, lodged in a space with a bunk bed for two others. Suitcases and books litter the floor. Damp clothing strung on the few hangers; backpacks and toiletries stacked high on small dressers. A shared bathroom next door with a door that won't pull completely shut.

It's been four years since my last workshop. A year isolated in hospital, another in recovery, then the pandemic, the following year—a bad Zoom connection. I wrestled with the decision to return in 2022. Was Hindman safe? I'm immunocompromised, protected by five COVID vaccines. A transplant recipient, I thought I could do it: masked, six steps back, meals at a distance outside. Wear the badge color indicating "No hugging. Stay back." I'd risk contact in class and while completing kitchen duty; I dared a virus to find me in the steam of the dishwasher.

In October 2018, in Washington State, I was diagnosed with cancer. Learned I would die unless a donor stepped up. Survival required a DNA/biological twin. Doctors started the hunt. Chemotherapy offered brief remission. The monthslong hunt continued. Magically, a stranger matched. He consented to donate. My transplant took place in March 2019. I was released home to recover in June of that year, but it would be one full year before my immune system was strong enough to risk a world full of people. My isolation ended in March 2020—finally safe enough to poke my head out. But *pandemic*, and the world shut down. Everyone tucked into the safety of home. My return to isolation felt less lonely.

The workshop has been a homecoming for me. I left Kentucky after college, went west. Wanted to see an ocean. Start a new life. But ever since leaving, I've been searching for a way to return.

Now I'm back at Hindman. This is the first workshop where I have been assigned a room in Stucky, the former trachoma clinic. It is the first time I've asked for a housing accommodation. Stucky is close to everything.

Monday night, asleep in my room around 3:30 a.m., I heard a small whimper. I rolled over.

A child's voice offered a loud complaint in my right ear.

"You moved."

I sat bolt upright. Grabbed the flashlight and cast the small circle about the room. My roommates, lumps under covers—I could hear their shallow breathing. No one awake.

I asked the room, "Did you say something?"

No answer. I brought my knees to my chest, kept a firm grip on my flashlight, as I waited for the morning to arrive. No more sleep.

"Is Stucky haunted?" I cast the question about the next morning. Shrugs.

"Sure, I've heard that before."

I asked a poet.

"Perhaps you heard your inner child speaking."

I thought about her wisdom all day.

I could see it. My inner child might mistake me for my mother. We look alike. Whenever as a child I needed attention, I'd go to her bed. She'd feign sleep. I'd reach over with small fingers and push open her eyelid, force her to see me. "*Mommy.*"

Trachoma clinic and eyes.

Pay attention.

"Move."

So many ways to move: location, position, pawn in a game, DNA swapped. My blood supply, now built from another's marrow. Transplanted, I'm a chimera—both male and female. If I bleed, the blood is his. You only find me in tissue or skin. I am the shell. His blood, the fuel that feeds what lies within.

I turned over in bed before the ghostly voiced complaint. Turned my back. Did I take up too much space? Did I come from too far away?

My first flight after the pandemic was to New York to meet my donor. We got caught up in Hurricane Ian. A hundred-year flood that left sandbags and water everywhere. My very next flight was to Hindman, Kentucky. What are the chances I would travel from a hundred-year flood to a thousand-year flood? No one could have predicted it.

On Wednesday, after class, a turtle crossed my path. "What do superstitions say about crossings?" I thought I should not risk it. I moved to the right. The turtle slowed to a stop, dead center, on the steep path. I needed a breath, so I sat down on the skirts of my raincoat, next to the turtle, and asked about its day. I did not notice the turtle was heading uphill. The off-and-on rain continued.

I looked into the eye of the turtle and wondered what the eye saw looking back. Am I changed in this body? Can the turtle see I'm an interloper, a West Coaster, a my-Kentucky-home leaver? Does it know I carry Scots-Irish marrow and literally, scientifically vetted Appalachian blood within me? *Do I belong?*

I live in a shell of my own.

I ask folks to keep their distance. But I get close to the turtle, get closer to it than any two-legged person around me. I want to photograph the eye. Turtles are forbidden. Immunocompromised, they hold great danger for me. No reptiles, no amphibians, no birds.

I take the photograph.

Turtles hold the world.

During my transplant, I developed a better appreciation for animals that hold over and sleep. Hibernation and emergence. A diapause of a season. A lost time, necessary to keep holy, a time to rest and live on.

Is shelter the place that I enter, or is it in a proverbial feeling that I carry in my heart center, wherever I roam?

Home, I believe, is to turtle. A verb. Grow a shell, plod along, and, when necessary, turn in.

The ancestors believe there is wisdom in journeys. Are you carrying what you need to survive? In a flood, turtles are displaced. They tuck into their shells and are cast into the current. When landed, they must start again. If upended, they must flip. A turtle on its back will suffocate within hours.

Some turtles live 150 years.

I talk in turtle because it's harder to speak about humans.

On Wednesday after dinner, I decided to miss the evening gathering.

I hadn't slept all week, not after that first ghostly night. I tried to sleep, but it wasn't long until I heard the rustling of voices: busy, full of inquiry and alarm. Like that of a hive that has been kicked, the hum became a roar and woke me.

I flipped the light switch—nothing. Grabbed my book light.

An emergency light illuminated the old staircase, with the wobbling handrail and irregular treads. I have warned everyone, "Beware at the top; don't trip." It's a long way down.

Downstairs, I saw the shadows cast by an outage light. We weren't in pitch black, but we weren't illuminated either. My night vision, terrible. I remembered the lantern I had packed last minute, hurried back to get it. Lantern features: an emergency beacon, cell phone charging port, exceptionally bright LED light. It had it all . . . except batteries.

I carried the lantern downstairs; ran into the poet again.

"I have a lantern. Does anyone have batteries?"

She looked at me as a kindergarten teacher regards a child, then reached over and patted my arm.

"You were close."

Writers were sitting crisscross all along the walls on the floor. Everyone piled under blankets.

"What's going on?"

Answers pelted: "The creek came up fast." "Evacuated." "We've lost cars." Folks started flowing in and out. Rain relentless, flashes of lightning struck all around.

I peered into the dark, watched the retreating forms. *This is not safe. Who is here? This is getting dangerous.*

Walking the dark rooms in Stucky, I started counting beating hearts. *How many? Why don't I know their names? Time to tally.* Moved back upstairs; loaded a backpack; added water bottles, enough for me and some extra; grabbed personal medications, a card noting next of kin.

Back downstairs, flashlighting windows with my phone, verifying they were unblocked: *Egress?* Two fire extinguishers; no first aid kits; bathtubs filled with water. I counted people; I counted pets. I was relieved to note that most were gathering toward the front. If the land slid, they'd have a better chance. I yelled out at folks wandering back into the storm.

"Come back—Stay!" I subtracted them from my tally.

I waited for the boom of a hill giving way. I tried to catch a glimpse of Troublesome between lightning strikes. Fumes, diesel, and gas mixed in the water, started to choke the folks on the porch.

Each lightning flash—in brief illumination—showed just enough horror to stop all conversation midsentence. Realizing the water on my back might be all I'd have for days, I cinched my backpack tight. I said to everyone passing, "Get your medications; put them on your body." I started handing water bottles out. "Keep this."

People were starting to panic. I got quiet, counted silently. I walked from top to bottom, front to back. Thirty-five to forty folks inside. Two dogs, one cat. I moved to the porch—heard a scream for help. In the dark, someone next to me said, "Did you hear that?" We listened; the cries stopped. I stayed on the porch, tuned for sounds of ground movement, pops, rumbles, trees crashing—feared the toe of the slope might give out. *Could this hill, like in the Oso slide, bury us alive?*[1]

Nothing to do until dawn. Only one person slept, snoring from the floor. I used the sound of snores to anchor another having a panic attack, grabbed her in a hug from behind. "Listen. Slow your breathing, Match, here. This moment. Stay. Don't run. Breathe."

Strangers walked in with their little dog. They never spoke. Tied their dog up in the bathroom, then wandered out. I never saw them again.

All had that long, hard, too-far stare. We were outside our bodies. Fear muted conversations. We were limited in response. *Endure.* It's all we had. No higher ground. No magic answer, no real supplies. Keep your shoes on. Stay, don't go. Daybreak is coming.

At the gray first light, three of us, without speaking, walked to the chapel, stood on the hill, and waited. Dawn came into focus; town was an apocalyptic scene.

We did not know that in a few hours we'd be provided a biblical parting of the debris in the water by the town locals. They would give us a path out.

Five vaccine boosters, masked, and yet . . .

I had started coughing that night. I blamed the smokers, the diesel, the damp, the asthma, the stress.

We moved fast, assigned carpools since many had lost cars. Piled extra passengers and baggage in. Get everyone out. Close, tight. No safe distancing. Pack in. Mud.

Tia Jensen

"How many can you take?"

We are liabilities to town now—no resources left.

"Leave while you can."

I carried the guilt of abandoning a community I'd grown to love. Leaving the cleanup behind. I coughed through my mask, and during the harrowing journey out, I gave everyone who rode in the car with me COVID.

I escaped to a friend's Airbnb. Home shifted again. In COVID isolation, I watched the calls for help through a computer screen.

Anger, grief, and frustration encased me. Four years I'd avoided COVID.

Turtle on its back. I struggled to breathe, pitiful until I poked my head out.

I moved fingers. Tapped out requests for donations. Shared stories. Broadcast messages. It is the tiniest of actions, this clickety-clack of fingers on keys. Words put to page: never adequate, never enough. It would be a week before I flew home.

There is no shell swapping. When danger comes, we can tuck inward and hide, or we can dig in and flip. Flipping alone is difficult. Believe in the kindness of strangers. Know that a life can be restored by another's gift. Hold onto hope. The earth is speaking. Answer her call to action. Keep moving. Turtle on! There is much work to do.

Volunteer blotting mud from the journal pages of May Stone and Katherine Pettit. © Melissa Helton

Left to right: Kristin Smith, Melissa Bond, and Jenny Williams preparing supper for the community. © Melissa Helton

Note

1. On the morning of March 22, 2014, the deadliest landslide in US history swallowed the town east of Oso, Washington. The slide destroyed forty-nine homes. At the time of the catastrophic event, I lived along the same mountain chain, in a geologically similar town, due south of the slide area. The hill in Oso gave way after a tremendous rain event. A factor in the collapse was the river running at the base of the slope, cutting through and destabilizing the hill. I knew a fast-moving erosive force like Troublesome could destabilize the toe of a slope. The community in Oso dug for four months. Forty-three casualties were pulled from the mud. The last victim was recovered on July 22, 2014.

Mandi Fugate Sheffel
What Water Can't Erase

AT WHAT POINT IS A HUNDRED-YEAR FLOOD NOT A HUNDRED-YEAR flood? At what point do we acknowledge the effects of climate change or its existence? When streets in the UK melt under record heat waves? When our coastlines continue to chip away because of rising water levels? When the frequency and severity of storms continue to increase exponentially? For me, it's when my home, eastern Kentucky, experiences a hundred-year flood once a year as we sit idly by and accept this as our new normal, as if there's nothing we can do to change it. However, what happened in eastern Kentucky on the night of July 28, 2022, is nothing short of monumental. We are not strangers to the effects of flash flooding after heavy rainfalls. It's something the region has battled for years. With our mountaintops removed, there's no place for the water to go, so it fills valleys and streams at a record pace, faster than the banks can hold. And we're left with homes sliding off mountains and water filling up living rooms.

    I watched the events of July 28, 2022, unfold from my hillside home for the week while attending the Appalachian Writers' Workshop. It's a week every year that is magical and sacred to all who attend. But it was clear to me that this was, in fact, not normal. I watched as some of my dearest friends, not from the area, lost their cars and spent all night watching the creek rise, wondering how they would get back home. But I was home; Knott County is where I grew up. And after the water receded and the roads were cleared, items were packed, and all the workshop participants found their way back home, I would have nowhere to go to escape the inconveniences of no cell service, power, or water. But I watched the reactions of others as they wondered how long we would be stranded and took pictures of the cars, tires, and garbage that filled Troublesome Creek. I was already thinking about the months ahead. Would this one get the national attention we would need to receive help from FEMA, or would it take days and weeks like the catastrophic floods of years past? The water would come up quickly but recede in the same fashion. But I could tell they were witnessing nothing they had ever seen before firsthand. And I could tell this one was worse than usual for us. Without contact with the outside world, I only later understood the enormity of what was happening all around us. The death toll climbed to forty-three. Six counties affected. Nineteen dead in Knott County. Relief efforts ramped up, and Appalachians did what Appalachians do, taking hits and helping neighbors.

I stood on the porch at Hindman. I paced and gave hugs. I watched as everyone drove off. I was left wondering where I would go next. My home in Jackson, Kentucky, forty-five miles northwest of Hindman, was inaccessible, with water over the road, and it could be days before I could get home. I knew this from experience after being trapped at my house for three days in March 2021 during a flood that brought record-high water levels to Jackson. My grandma's house in Hindman was a total loss, and now she's living with my mom. So I decided to go to Hazard and assess the damage at the local bookstore I own downtown. This became part of my routine after I opened the bookstore in January 2020 and experienced two feet of water in the store a week later. I drove and thought, thinking about these floods, one a year for the past three years, and those are just the ones that have had a direct impact on me. This doesn't account for the countless others often localized in the region.

As the devastation hit social media, it was scene after unbearable scene. Two of our most sacred spaces, Appalshop in Whitesburg and the Hindman Settlement School in Hindman, have lost years of archives. Appalshop was founded in 1969 as the Appalachian Film Workshop, originally a project of the US government's War on Poverty. The people of eastern Kentucky have used Appalshop as a resource to preserve our rich cultural history. Years of pride and preservation. Years of work to preserve our existence. A history of people forgotten and a time erased by most. A past we choose to document and preserve. We tell our story, not the narrative pushed by many who visit from the outside. Is that what we have become, a place on the map that time and water can erase? With these documents gone, how will our children know the importance of who we are and what we have overcome? How do we move forward with a failing infrastructure that continues to become increasingly fragile with every rising tide?

In the weeks that followed, I found myself restless, and I left home most days with no plan for what to do or how to help. The work seemed insurmountable. Saturday after the flood back in Hindman, the Settlement School had been turned into a makeshift donation hub while also housing displaced flood victims. I helped wash and dry photos from the archives. These photos tell the history of a place that has become more than just a school but a home for our cultural heritage. Some days I loaded my car with supplies, water, Clorox, gloves, and snacks. I picked a place, any place, because everywhere I looked, people were living in tents and working against the clock to save what they could and remove the flood mud from their homes. As they worked long hours in the heat of August without power, without water, the stench of flood mud lingered in the air. I'd ask them what they needed, and often the response was, "We don't need anything. My neighbor down the road could use it worse than me," from people who had been without a hot meal or shower for days. I provided what I could, which sometimes meant just a listening ear. Everyone had a story, and they were desperate to be heard.

There were days when I mucked houses for strangers and family. Rubber boots, crowbars, gloves, and empathy. East Kentucky Mutual Aid, a local nonprofit, became a saving grace for a lot of families by taking donations and putting money straight into the hands of flood victims. Folks who didn't have the time or resources to wait on FEMA. On two separate occasions, I was given an envelope of cash and told to give $200 to each household in the hardest-hit areas of my county. This small gesture profoundly affected me, as a recovering addict of seventeen years. I broke down. There was a time when no one would have trusted me with this task, and rightly so. In the past, I would have used this money for drugs. During this time, I thought a lot about the addicts still in active addiction or those new to recovery. Periods of forced isolation like those caused by COVID and flooding can cause a rise in relapse rates.[1]

Signs on the road say "Recovery Is Possible." Today these signs have a double meaning: recovery is possible. We know it because we continue to do it. We are constantly recovering from industries that recognize the potential for profit that comes at the cost of our land and its people.

People were so grateful.

I carry their stories with me. Stories of bodies hanging in trees while emergency workers struggled to find access. A man who stayed in his trailer and rode the rising tide until it came to a stopping point. Everyone's story was the same: "It was like a wall of water," or "This felt like an inland tsunami." Creeks that generally flow no more than six inches deep became raging rivers.

Some say it was a mine disaster.

What I saw in those hollers is something hard to put into words. Houses were picked up and carried miles down the creek, cars and bridges wrapped around trees, and debris hung from power lines. Lives and hard-earned possessions collected like trash in a dump on the creek banks. But there's hope, there was always hope, and resiliency remains. That's something the water can't wash away. But people are tired of being resilient, tired of swimming upstream, and tired of being the ones who must recover.

A hundred-year flood has a 1 percent chance of being equaled or exceeded within a year. The flooding that occurred in 2020, 2021, and 2022 could all fall into this category. But why here? Why now? What we are experiencing is unlike anything I have seen in eastern Kentucky. Beyond the need for explanation is a feeling of hopelessness for this region I love. I see people who were beginning to rebuild a life after the flood of 2021 knocked flat again. But I also see a community putting aside everything that divides us as a nation to save lives. I have hope that we will persevere and work to make a life in this place we call home. Many of us have been here for generations. We will do the work to save our precious archives, our only remaining connection to the past. After we clean up, after we rebuild, we must address how climate change and extractive industries are erasing our future.

Note

1. For more information about the opioid crisis, see the work of Beth Macy (*Dopesick* and *Raising Lazarus*) and Sam Quinones (*Dreamland* and *The Least of*

Us). For a humanizing fictional account of the crisis in Appalachia, see Barbara Kingsolver's Pulitzer-winning *Demon Copperhead*. Helpful articles "COVID-19 and Addiction," by Dubey et al., and "Understanding Postdisaster Substance Use and Psychological Distress Using Concepts from the Self-Medication Hypothesis and Social Cognitive Theory," by Alexander and Ward, can be found on the National Institutes of Health's National Library of Medicine website.

Storage room on the ground floor of the Mullins Center. © Melissa Helton

Darnell Arnoult
Knowing

River breathing your
water's dark body.
Wild premonitions.

G. Akers
Stag: A Story

If the world has taught me anything, it is that in the darkest of days we kneel at the feet of the world around us. We often ignore it—pretend as if the sky does not know our names, so we scream at it. Cursing and throwing stones as if that will change anything.

I am trying to capture this with my eyes—brown and wide, staring into the headlights of a car that does not seem to stop.

I have heard that they often never do, not daring to believe they could kneel to nature until you are smashed on the hood of their car, antlers through the windshield on a chilly night.

The best, though, is when the sky screams back. Deafening us and dragging our feeble ideas through muck and mud because occasionally we need to be reminded just how bright the sun burns.

I do not know why I was drawn so close to the sun that day or why, when presented with pages and pages of manuscripts and history, I felt as if maybe I could dare to understand it. I wanted to.

I wanted to know what the birds sang like before the ground was tilled. Turned upside down for people and food that suck from the soil until it is dusty and thin.

I wanted to give up all of myself if it meant that the sun would be mine. That I could hold its burning embers between my fingers until it melted my skin and eyes. I like to think that all that would survive the scalding would be my glittering, golden self—immortalized by the heat. Gold-bathed bones clinking to the rocks, only to be found years later under the rush of a stream.

TROUBLESOME RISING

"You can't remember things if you don't try to. Just forget what you don't want to know." That was your advice, and yet I didn't listen. Because I know that even in destruction, it was destruction *I* caused, destruction I endorsed.

In the dark hours after dawn, I slide my way down to the creek, letting the soggy ground take me farther into the deepest part of the land. I can smell it here—the scent of living clogging up my nostrils so much differently than the perfumes they use.

Yet I feel almost akin to the feeling it possesses—knowing that it acknowledges just as well as I do that there is no forgetting; there is no pride. There is only power. There is only what holds itself over us and what we hold ourselves up to.

There is only the trembling in my legs as I dip my head down into the water and drink, open to the arrows of a world that has never known the earth.

Pauletta Hansel

Aerial View of Catastrophic Flooding in Eastern Kentucky

Quicksand, Bulan, Neon, Hiner, Martin, Fisty.
This is our place in Hueysville.
This was my mother's house before she passed.
Samantha's sister's house is by that blue bridge.
Anyone know anything about Fugate's Fork Road?
Stringtown, Ajax, Isom, Pinetop, Dwarf.
This is my cousin's house.
My mamaw's house is on the left.
That bridge is about eight feet above
where the creek's supposed to be.
Isn't this Mary's house?
This is the mouth of our hollow,
the red arrow was our road in.
Nix Branch, Jakes Branch, Trot.
If you zoom in to where the white car hood is,
my home is there.
Rowdy, Wayland, Noble's Landing, Cowan Creek.
OMG that is Pigeon Roost.
Hindman, Buckhorn, Chavis, Krypton, Garrett,
over toward Pound.
Y'all, this is my hometown.
This little tree, and God, kept us alive this morning.
My daughter swam with her dog to a neighboring rooftop.
Does anyone know about Kite, Kentucky?

TROUBLESOME RISING

Caney, Possum, Ary, Lost Creek, Hardburly, Trace.
Dad and my nephew are neck deep
they need help
please.
Are you all safe?
We lost the farm animals and five cats.
Lost my chain saws so I can't even work.
You need to understand the nature of the topography.
Add to that strip mining, climate change, political neglect.
We have lost everything
again.
We have warm beds, clothes, and toiletries available.
We have hot showers and food.
Anyone trapped in downtown Whitesburg is welcome to come.
We need help and I'm willing to help anyone
in the same shape we are.
Your prayers are good
but we need to get federal and state assistance ASAP.
Don't cry for Appalachia, work for change however you can!
Let's use the internet to tell our story.
Thank you for posting.
Much love and many blessings to you all
from what's left.

Randi Ward

Aquarius

The signs
were there:
in the air,
in the creek—
I'll bear it,
in my blood,
indefinitely.

Melissa Helton
Chain of Custody

Humane Society employees
off-load pallets of dog food
for our flood survivor donation
center two weeks after the flash.

A black rat snake, fallen out
of their wheel well or dropped out
of their bed, slowly thrashes
under the truck, a red-purple

bulge in its side. I see its beating
little heart the size of a thistle
blossom, a toddler's big toe,
a garden strawberry, a grown

man's thumbnail, a spice drop,
half a square of Turkish delight.
It curls into itself beside
the back tire, slowly like a groan.

Its heart pulses out in the air.
My kid gets a shovel and a box,
eases it up and in, whispering
a shushing sound of comfort.

Melissa Helton

If I were brave, I'd chop that shovel
edge down on its neck and end
the writhing immediately. I stand,
head full of static, watching that

beating heart, beating out in the August
sunshine, the air caressing what it
should never touch, brushing
against this pulse that should

be secret and dark. My kid carries
the box, reverent, to the creek
for a water death and burial,
cooing, *Oh pobrecito, shhh*.

Tomorrow I will stand
at the bank, over the most
likely dead snapping turtle nest
we had marked with an orange

traffic cone all summer, protection
from feet and lawnmowers,
the calendar ticking closer, square
by square, to August 8, predicted

hatch day. After twenty-four hours under
cold, toxic floodwater, I will try not to
hope but also won't be able to help it.
I will think about the scaly, most likely

dead babies in little egg graves
buried to incubate a birth that won't come.
I will think about their water death
and all the other water deaths

in these counties. I will look into
the quiet, clear bottom, a scant six inches

TROUBLESOME RISING

deep, see momma snapper over
the black rat snake no longer writhing,

heart no longer pulsing. I will ask
live momma snapper and dead
black rat snake if the babies in their eggs
are not-yet-born or never-will-be-born.

The only response we can expect
is the trickle of Troublesome,
that water which flooded us all out
and brought the need for these

donations, brought this truck,
and this pulsing-not-pulsing snake,
and this nest of probably drowned eggs,
and this nest-building momma turtle,

and me in the first place, here,
to bury this in my heart, and now yours.

Pamela Hirschler
With What Remains

Muddy water still cascades down the mountain,
a froth of current eddies and pulls at the bank,
plastic bottles snag in the bend, a high chair
dangles from an oak with toilet-paper branches.

How many bags will volunteers haul away,
how many cans will be traded for change,
how many families won't worry about river trash
while they shovel mud from their living rooms?

But the bloodroot and bluebells still emerge,
the redbud blooms, and the wren twitters at its mate,
pushes a tattered drinking straw into the nest,
and with a thin bill, weaves it home.

You can know a place
by what the flood leaves behind.

III | These Sunken, Unpeopled Streets

Will Anderson speaks to military personnel gathering data for FEMA deployment. © Melissa Helton

Kelli Hansel Haywood
Backwash

1

Water got over the road with the rain last night and again this morning. As soon as I got muffins in the oven for breakfast, the electricity went out again. I tried to ignore the rain. Tried not to stand in the window and watch the creek.

I see the outage is due to the power company coming back to string the new poles they've placed. Not the storm, this time. Their workday here was unannounced to us in the holler. They keep getting interrupted by rain. Their trucks block the entire road. No way in or out.

A neighbor has to park her car in my yard and head up the holler on foot. She's trying to reach a family member who requires oxygen and needs help because the equipment is electric. She knocked on my door to ask permission. Another woman comes and demands they move the trucks to let her through. Says she doesn't give a shit, she'll go to jail. Same reason—family member on oxygen, now smothering. Second woman is angry. Understandably. Anger manifests from a lot of sources, most of which are not originally anger. Anger is easier for us to deal with than some things. They move the trucks.

Again, the theme—I can't breathe. Smothering. Waiting.

No cell service here. No internet without electricity. Landline works. Today. Rain pauses.

Another knock on the door. I startle.

An old friend stands outside. He doesn't live here anymore but is trying to help as he can. He has forty single-gallon jugs of water he is delivering for distribution to the church down the road. Some folks are looking at a month without running water. Maybe more. I help him unload them here. I put a few back for my sister. She's one of those waiting.

I notice that my left arm is quivering, right before my friend asks how I am. I don't know how I am. Instead of answering his question, I start telling him some of the experiences of the last week. He's worried about being able to get back out of my yard as more uninformed neighbors are trying to drive out of the holler. He leaves.

Many hollers are one-lane roads with no shoulder. Driving here, we learn to maneuver vehicles pretty well. Lots of tight squeezes. Sometimes you can't get through.

My arm still shakes. Not sure when the lights will come back on. At least it's not raining right now.

Later in the evening, someone puts their car directly into the creek bed over one of the highest embankments. I'm trying to take my daughter to see a friend. She needs a shower. It looks like they drove into the creek as if it were a road and they were trying to drive through the culvert. There's a crowd and state police. My daughter grows too anxious to attempt to leave. We go back home.

2

Somewhere around the second day, when the waters began folding back, I noticed the hummingbirds. Hummingbirds hovering around the creek bank looking for the different blooms they had drunk from days before. Hovering, circling around and back to hover again, finding nothing but mud and rock and muddy water where their life once lived.

3

Maybe it was the fourth day that I noticed the kingfisher. I know it was a day in the middle. The kingfisher fitfully hurried back and forth between the mountains that wrap my holler, across my yard. It had been almost a week since I had told my writer friend and mentor Jim Minick, an avid birder, that I wasn't sure if I had ever seen a kingfisher around.

I had just received the gift of a feather from a blue heron I had spotted walking across the campus of the Hindman Settlement School. As I showed Jim my treasure, we talked about birds.

"If you have water near, you've seen them," he said. Jim pulled up a picture on his phone. I had been mistaking them for downy woodpeckers.

I recognized, then, the bird that always sits in the Upper Bottom of Whitesburg on a power line that crosses the North Fork of the Kentucky River where it hugs the front edge of the bottom. The Upper Bottom, where so many of the nice homes were, was washed out in the flood.

Is the power line still there? The downy woodpecker turned kingfisher?

The kingfisher fitfully crossing my yard again and again made me picture Jim's flushed and joyful face as we talked just those days back.

4

The disconnection felt like sinking. Sinking beneath the waterlogged ground. The swells of water, still pushing their way down the carved-out creek bed, now three feet wider, with a consistent low roar.

As a little girl, I spent entire days up the mountain, among the big rocks, along the old cow paths and logging roads, watching for copperheads like Dad taught me and scanning the ground for treasure left by nature—shiny rocks, dried bones, and feathers. Entire days where not even my mother could find me if she wanted to. Entire days not existing in sight of anyone but myself and the earth. She'll love you and swallow you whole.

A sinking in the pit of the stomach. Like losing. A dark dropping out of a world of one-click communication, pictures of puppies and food, socialization intangible, and somehow it made me . . . *me*, uncomfortable to be gone from it. Sitting in the yard under the stars, watching with that sinking feeling for bars to appear on my phone. Any sign of getting a text through to someone. How long is it before your absence on the screen means you don't exist? Was anyone looking for me?

It is all maya anyway. Technology. That we exist. That I, a mountain woman up this Kentucky holler, uses Sanskrit words and reads about Hindu philosophy. When it's gone, I am still what I am. That girl in the woods. A member of the lost, the got lost, the hiding, the resilient, the make-do, the God fearing, the left alone. The mountain woman.

Me and the night sky full of stars, penned in and held by mountains on all sides. My phone buzzed. I felt my heart race in my chest as I looked down.

Hey, it's . . .

First contact. Someone I have never met in person. Someone in Canada. Someone I had never shared my cell number with. Cared. Searched for me. Noticed I was gone. Found the news. Searched. Spent time looking for someone they have never met in person. Found me.

I will love him forever.

It was an hour before it occurred to me that international texts may cost more and that I would have to go back to the dark.

5

I know it was Day 6 because the electricity was restored that evening for the first time.

A tiny brown-and-yellow butterfly came to where I lay on the cool concrete of the shady end of my porch. My veins bulged with pooling blood. The heat and stress had brought my symptoms of POTS (postural orthostatic tachycardia syndrome) into play. I couldn't stand more than five minutes at a time. I couldn't move my heart rate above fifty-six.

The butterfly was dying. I could tell by the drunken way it moved. The flitting of wings with no flight. I reached out my hand. It climbed onto my fingers and fell off. Trying again, it lifted itself from my fingers for one last taste of life in the air before it rested again beside me on the cool concrete. The air, too heavy for both of us. As I slept there, the butterfly died.

6

"I'm worried about my mom and them," my neighbor's partner said, as my neighbor walked down what was left of my yard, looking at the roaring deluge of water that covered a huge chunk of the yard and the road. He works for the highway department. I could tell by his body language that it wasn't good. They'd already told us that the hill behind our homes had slid off, hitting the back of their house. None of us had a way to understand what our neighbors down the holler and throughout Letcher County were experiencing. There's no reference point for a thousand-year flood. Fortunate enough to be at the headwaters, all we knew was this was a type of fortunate that made you wonder about things like luck, hard work, and progress.

I could barely hear her soft voice over the roaring water. "I don't know how we'd even get to them. I can't call," she said.

It would be days.

I had woken briefly in the night. I could hear the pounding rain. I expected the power to be out soon. It goes out with most storms in this holler. The internet too. I had left the Hindman Settlement School's Appalachian Writers' Workshop earlier in the evening, when the rain began, telling people as I left not to celebrate the rain for keeping things cool in the near end of July. We didn't need any more rain. There had already been two small flash floods over a period of a few weeks, the last of which had waterlogged my minivan as I tried to get up my mother's holler to pick up my daughters the week prior. It was in the mechanic's garage.

We didn't need more rain.

When I woke up to the faint light of morning, it was the kind of quiet you can only find without the buzzing of electricity. A stillness of endings and grand pauses. Until my ears picked up a new sound. Rushing. Roaring. Roaring. Not rain. Monumental.

I went to the living room to look out into the front yard.

It didn't ask for a pause. It demanded.

Shelly Jones
Floods Make Fertile Ground

Floods make fertile grounds, or so I learned from the grumpy Mr. Brookbank in seventh-grade social studies class. Sitting like a commander behind his desk, his arms folded over his belly, he instructed us to take out our current-events notebook every Wednesday. From our *Weekly Readers*, we'd distill major world news into three tidy columns per loose-leaf page. Holding a wooden ruler steady with one hand, its metal edge sharp against my pencil, I zipped two straight, parallel lines down my paper and labeled the three columns "When," "Who and What," and "Where." Mr. Brookbank ignored the how and the why.

The third column was always my favorite: "Where." Here we traced the outlines of states, of nations. Oceans, rivers, mountain ranges, borders. Egypt, no bigger than a half-dollar coin, crisp on the page. My mechanical pencil traced zigs and zags, veins of charcoal. The Nile delta snaking and branching in the tight column set against pale blue lines on the soft white paper. Some boundaries are decided by nature; some are created by greed. When Mr. Brookbank shared important events with us, he never said much about the people or lives affected. It was just the basic facts, no feelings. Thursday, July 28, 2022—the day I witnessed catastrophic flooding in eastern Kentucky—couldn't easily fit into three tidy columns. The when and the what are both fairly straightforward. The who and the where? Much more nuanced.

Wednesday afternoon, the day before the flood, I sat on the front porch of the May Stone Gathering Place with my teacher for the week at the Appalachian Writers' Workshop, poet Nickole Brown. A poetry workshop conversation. Two notes scribbled in my notebook from that meeting, Nickole's learned wisdom from her teachers, passed along. Earlier the same week, poet George Ella Lyon had shared that as a writer, she learned that she should "sing with the voice you have, not with the voice you want." She continued: "It took me a long time to come home to the voices that made me." My leverage point in the conversation with Nickole was about that very thing—using my true, real, authentic voice.

I left the conversation, tears rolling down my face, knowing what I had to do. Walking past rain-soaked hydrangeas and the massive bigleaf magnolia draping over Uncle Sol's Cabin, Robert Gipe, arriving to visit for the rest of the conference, stood there holding a bag of Grippo's. "Hey there!" he drawled, noticing my face. "Looks like you got your tuition's worth." We stood and chatted for a few minutes. A classmate, Jim Minick, approached and started talking about vultures vomiting in self-defense. Our homework assignment that evening was to write about something disgusting, and the subject of vomit had come up. He explained that vultures sting potential predators' eyes with their acidic puke while also lightening their own load so they can fly away more easily.

Gipe turned to me again and said, "So—what were you crying about?" I responded, "Well, poetry conference. All the feels. You know how it is." He laughed and agreed, not seeming completely satisfied with the lack of transparency in my answer but not pressing further.

I hung around for the rest of the afternoon and evening for the faculty readers, Meredith McCarroll and Carter Sickels. On Sunday night, we'd all sat on the banks of Troublesome Creek and watched the film based on Carter's powerful novel, *The Evening Hour*. And there wasn't a dry eye in the Great Hall after we heard Meredith's essay about her mother. A line she shared still haunts me: "I'm writing to you dead to keep you alive." After the readings, knowing Thursday night is normally a late night at the workshop, I decided to skip trivia night since I still had poetry homework to finish and wanted to be ready for the last full day's festivities. I said my good-byes for the evening and drove back to the Hampton Inn in Hazard in the rain.

I'm a heavy sleeper. After ignoring the middle-of-the-night flash flood alerts on my phone, I awakened in the morning to messages of folks from the workshop checking on one another but wasn't aware enough yet to understand why. I gathered that the campus had flooded somewhat, which didn't surprise me after the warnings Josh Mullins had given us earlier in the week about the lower-level parking lots. I naively wondered if classes would be canceled for the day and if I should go ahead and get up to take a shower. Pondering whether to go to the lobby to make a waffle for breakfast or start some coffee in the room, I stumbled to the window, pulled back the curtain, and rubbed my eyes. The volume of water covering half the parking lot was astonishing. To say it looked like a small lake doesn't suffice. I went back to my phone and felt a heaviness. It had rained so much overnight. Things were worse than I had realized.

More messages poured through and were posted on the Hindman Writers' private Facebook page. Stories of poets and storytellers escaping floodwaters, huddling on the Stucky porch in flashing thunder. Stories of cars being moved, and moved, and moved again to escape the rising Troublesome Creek. As the sun came up and folks began to share more details and videos on social media, the picture came more clearly into focus. Friends' vehicles on Hindman's campus were flooded in places where my car had been parked just hours before, places we thought to be high ground, out of Troublesome's reach. People were trapped on campus: the Troublesome Creek too high, the bridge not passable, the town of Hindman underwater. So many roads impassable. Untold lives lost and folks in need of rescue. I quickly got dressed so I could leave my room and see what was happening.

Outside in the hallway, my sneakers squished onto sopping-wet carpet. The hotel lobby was filled with people stunned and milling: state police in camo murmuring of 2,300 underwater in Whitesburg, women still in their nightgowns sleeping on couches after moving during the middle of the night when their rooms took on water, traveling doctors in scrubs checking out and heading home. Others of us staggering around, unsure of what to do. A local couple spoke in low tones to their son who had worked all night at the front desk. They shared with me, just as a matter of fact, that their neighbor had washed away and they couldn't find her. She was there, and then she wasn't. Everyone in their holler had lost everything. And it wasn't just their holler. Outside, half the Food City parking lot was filled with several feet of water, and a huge chunk of the soil on the bank behind the hotel had given way, soil and rock and cinder blocks strewn down the hill and across the road like a pile of muddy Legos.

What could I do? What should I do? What did people need in emergencies? Clean water. I knew that would be something people could use. I decided to drive across the not-flooded side of the parking lot to Food City, buy as much water as I could haul, and take it to campus when the roads were passable. I just wanted to *do* something. Anything at all. But then I began to wonder: "Do I need to just get the hell out of here? Am I in the way? How long do I stick around and try to help?" I drove my car back to the hotel and asked the young man at the front desk, who had been fighting the couple of inches of mud and water flooding the back of the hotel all night, for his thoughts. Did they need my room? He said, "Ma'am, I'm not going to tell you what to do, but we are receiving multiple calls from locals looking for places to stay. Folks who have lost everything. Even hospice is calling, asking if we have rooms." I had my answer. I gave the carload of water away to the couple whose friend had washed away, loaded my belongings, told the front desk that I was leaving early, and asked them to give my room away to someone else. Once the all clear came from Hindman that waters were down, everyone on campus was able to leave, and everyone had rides home, I knew it was time to go.

Trying to return to Louisville the way I had come was the wrong move. The images I saw through my car windows were in the *New York Times* the next day: buckled and washed-away chunks of the highway; women huddled in blankets beside pickup trucks along the side of the road; dogs roaming, confused; school buses tossed into the sides of buildings. A washing machine sat on a bridge as though it had been placed there on purpose. Rooftops, stripped of shingles, straddled guardrails. The steps to somebody's trailer were pitched sideways on the bank beside a dresser turned upside down. Trees bent at their bases, knocked over, stripped of leaves. I had never seen such devastation. Evidence of water's power, the helplessness of those in its path, the destruction of homes and lives and livelihoods all happening in a matter of hours.

I kept driving, terrified, until I couldn't. A group of men in coveralls blocked the road with their pickups, yellow lights flashing. I rolled down my window and they asked me, "Where you trying to go?" I said, "Louisville." They told me the only way to Louisville was to go back the way I'd come, all the way to Hazard, get on the Hal Rogers Parkway, and head west all the way to I-75 and north from there. I thanked them and turned around.

Rain began to fall again. Muddy water ran almost parallel to the road. My fear grew. Would this river spill out of its banks again? The buckled road could give way at any moment. Would I make it to the Hal Rogers Parkway? I felt such shame and guilt at leaving. Escaping. Appalachia is my home and yet not my home. I've been an Appalachian expat since 2000. I had a safe house and a clean, warm, dry bed awaiting me just a few hours away. Yet in front of me, buckled slabs of asphalt as big as my entire vehicle lay strewn about the mud-covered road. I drove around them, gripping my steering wheel tightly, the sound of the windshield wipers a strange and rhythmic comfort.

Pickup trucks flashing yellow emergency lights and pulling airboats and generators headed toward Hazard and beyond. Help was on its way. The farther west I drove, the drier, sunnier, and more normal things seemed. My body still tensed and trembled from the adrenaline of witnessing such devastation. By 12:25 p.m., I texted a friend: "George Ella is two cars behind me at a stoplight in London. It's a balm to know everyone from the workshop is safe." Earlier that week, this same incredible human and poet I had just passed in her little sedan, her hands gripping the wheels at ten and two, had laughed while storytelling over lunch. She talked about how she had seen one of her poems in a textbook and how she, the poet who *wrote* the poem, couldn't answer a single question the textbook asked. She said it was akin to someone eating a meal and then writing a chemical analysis of the specific ingredients. She scolded, "The first question someone should ask themselves after reading a poem is 'How did it make you feel?'"

Along I-75 North, cars zoomed past, and truckers distractedly swerved in and out of their lanes on the dry, fully intact highway. I needed another dose of Appalachia before going back to Louisville, and the Kentucky Artisan Center in Berea, a place I frequent when traveling, was my next stop. I walked in and sat down in the quiet café to take some breaths and reconnect my mind and body, to feel grounded. To remind myself that I was safe and I was going to be all right. I noticed another writing workshop refugee, Loren Crawford, walking into the building. And I was so grateful. Less than twenty-four hours before, I had snapped a photo of her "Puck the Faetriarchy" Shakespearean-themed T-shirt, which had been her trivia team's namesake. We hugged and expressed mutual gratitude for our safety, ordered some lunch, shared stories of the hell we had seen trying to leave, and made plans to be in touch soon. Already a sense of guilt at leaving infused the conversation—a sense of what we had lost, yet a knowledge that others had lost so much more. We wished each other safe journeys home.

Walking to my car, I noticed a spray of mud the color of caramel on the sides and back hatch of my white SUV. It would be weeks before I could make myself run it through a car wash. The mud, a tangible witness, an artifact of what I had seen, somehow connected me. This tie to place—to the hills and hollers and creeks and streams, to Appalachia—runs several generations deep in my DNA. Unlike many Hindman friends, I couldn't go back and help muck houses, preserve archives, or distribute supplies. My work starting a school year for teachers and families was ramping up in Louisville, and I could not leave. Instead, I did my best to help from afar through donation drives, supply drives, and events supporting the relief effort. The help poured in, though, and collectively, the work continues. Forty-four lives were lost in this flood—so many fewer

Staff and volunteers unload donations of drinking water. © Melissa Helton

than first feared—and there were also tremendous losses of infrastructure and housing and livelihoods, of culture and history.

Back to my seventh-grade current-events notebook and Mr. Brookbank's third column, "Where": there's no way a space the size of a half dollar could hold the map of this region, this *where*. The storytellers and artists who know and love Appalachia best won't look away but rather will examine, cherish, and claim the truths that live there, even the ones that need upending. The Hindman writers will unpack and unzip and expand upon this flood and its impact for generations, gathering along the fertile, sacred banks of the Troublesome.

Monic Ductan
The Gift Horse: A Story

My cousin Eugenia swears that she has the gift of second sight. I don't believe much in the supernatural, but I will say that Eugenia called my mama's death before it happened, said Mama would die by the side of the road and her body would be mangled in a car crash. A week later, Mama's car was T-boned by a dump truck. Eugenia has called other things in our family's lives, too. She knew when our cousins were pregnant before they did. Eugenia knows too much. Sometimes she will sit staring, her black eyes unblinking, and when she does that, I can tell she's seeing something she doesn't want to see.

As we walked down the gravel road leading away from my house, Eugenia said, "If the creek rises tonight, I'll wake you up, Mandy."

"Nobody is calling for a flood," I said, trying to read the look on her face. Had she seen a flood?

I had just moved into the house that morning, and Bailey, the man who'd sold me the property, had left his horse fenced in the backyard. I'd called all day to speak with him, and he'd not answered me. Now Eugenia and I were walking the horse up the dirt road to my neighbor's house. The plan was to give the horse to my neighbor. I wanted the animal gone by that night so I wouldn't have to march him into my old rickety-looking barn surrounded by tall weeds that needed bushhogging.

"Damn Bailey for sticking you with this danged horse," Eugenia said. Her black hair had turned frizzy because of the rain. Eugenia was a short, quick, dark-brown woman who spoke fast, walked fast, and had a quick temper.

I silently cursed Bailey, too. I didn't know the first thing about looking after a horse, and he'd promised to have the horse carted off before I moved in.

It started to rain again, and the dirt road turned muddy. I pulled my slicker's hood tighter over my head. Eugenia held the horse's reins. The boots I wore weren't comfortable, and clumps of mud and pebbles kept lodging themselves under my feet.

Eugenia sighed. "You sure there's a house back in here?"

"You can see part of it through the trees," I pointed out as I gestured with my hand.

The horse snorted and sneezed.

We had to walk uphill a ways and then downhill again before the house came into plainer view. It was a two-story, painted pale blue.

Eugenia let out a groan, and I looked over at her. She squinted at something in front of us. When I followed her eyes, I saw that a Confederate flag, sagging with rainwater, hugged the trunk of an oak tree beside the house.

"Goddamn," Eugenia said, shaking her head from side to side.

She'd warned me that I would be the only Black person in my new neighborhood, which seemed true given what we'd seen so far.

As we got closer to the house, I noticed a towheaded boy playing on the porch. He looked about the size of the kindergarteners I teach. When he saw us, he jumped up and ran into the house.

Eugenia took me by the arm. "Let's get outta here." She turned back and started to drag both me and the horse with her.

I pulled my arm from her grasp. "What are we going to do with this horse? I don't want to see after him," I said. "And it takes time to place an ad to sell him. It could be days before we find a buyer."

A man came out onto the porch, but Eugenia had started back down the path away from the house. "Wait," I told her. I took the reins from her, turned the animal around, and led him up to the porch.

"Good evening," I said to the man. "I just bought the farm up the lane here, and Mr. Bailey abandoned this horse in my pasture. I got no way of taking care of him, and I thought you might like to have him."

The rain had eased up some, but it still dribbled like a leaky faucet. The man squinted down at me from his spot at the top of the porch steps. Something told me that he was in his thirties like me, but he had a lot of sun damage on his face—cheeks flushed red and freckles on his cheeks and hands.

"How much?" the man asked.

"Oh, no. Nothing. It's no charge. I'm trying to give him a good home. He really just needs him a warm, dry barn right now. Maybe Bailey will claim him tomorrow, but that'll be too late. Far as I'm concerned, he's yours if you'll take him off my hands." Because the man was looking at me like he thought the horse was counterfeit or something, I rushed on. "I'm not a farmer. I don't know how to see about a horse. I don't want him to suffer." I jerked my thumb over my shoulder toward the man's barn. "I see you have a barn, so I thought you might want another horse." Trying to explain myself to this man made me realize just how dumb my plan was. Wasn't I supposed to have some sort of papers drawn up to give the horse to him?

The man came down the steps and opened the horse's mouth. He peered into it, tilting the horse's head this way and that. "I can have a vet look him over in the morning," he said.

"If there's some papers to fill out or something, we can handle that next week," I said. "Thank you."

I turned to go.

"Hey," the man said.

I turned back.

"How much?"

"No charge."

I hurried down the road after Eugenia.

"Hey!" he called again, but I just waved and kept slogging through the mud. Finally, I heard him shout his thanks, which echoed through the valley.

That night I felt antsy because of Eugenia's prediction that the creek might rise. Just when I was nodding off, she tapped on my door and stuck her head in. "You up?"

I groaned. "No, I'm asleep."

"Mandy," she said to me. "We gotta get out."

"*What?*"

"Get your shoes on, girl. We can go stay at Jason's tonight."

Jason is my brother, and he and my cousin Sam were the ones who'd be delivering the rest of my furniture the next morning. Jason lives over a hundred miles away, and the last thing I wanted was to drive for almost two hours in the dark while I was groggy and sleepy.

But Eugenia was already pulling the covers off me. She tossed my robe onto the foot of the bed.

Outside the house, a horse whinnied in the distance, and I wondered if my gift horse was calling to us from the neighbor's barn. As Eugenia backed us out of the driveway in my truck, I looked at the beautiful old house I'd bought. It was a two-story building that gleamed ultrawhite in the moonlight. The land was one reason I had bought this place. Eight acres. Enough to raise some chickens and a milk cow, the real estate agent had told me, obviously giving me his sales pitch. But I had no desire to farm the land. I just liked the wide-open space.

Two days later, the flood came. It washed out our valley, and I prayed my house would somehow still miraculously be there, though I doubted it. We saw the footage on the evening news from the safety of Jason's couch. Knott County was one of the hardest-hit places, and my house sat in the valley, a perfect spot for flooding. Thank God for Eugenia's prediction. Otherwise, we may have drowned. Still, I wondered if we could salvage anything.

It took several weeks before the main roads opened again, and when we finally did get into our county, they said the valley roads were still washed out.

My property was insured, but I was told I needed flood insurance if my house had been one of those washed away. I didn't know if it had washed away. I couldn't even get back into the valley to check it out.

What about my neighbors down in the holler, the ones I gave the horse to? Eugenia scoffed at me when I mentioned them.

"They're hateful," she told me, and she reminded me of the Confederate flag on their tree.

"Let's go see if the school is still there," I said.

Eugenia and I both taught at the same elementary school. Driving to the school, we passed creeks that carried wood and debris. A bicycle lay on the roadside, as if it had been haphazardly dropped there by a child.

The school didn't sit as far down in the valley as my house did, so most of the building was salvaged. Still, I felt lucky to have my gum-rubber boots on as we walked through the school cafeteria. Water damage and mold had left their marks all over the baseboards and floors. Some areas were inaccessible because of the inches of mud caked on the floors. I plopped down on top of a table and cried.

Eugenia put her arms around me and reminded me of the help the government had promised.

"You believe those lying-ass *politicians*? You are the very one always criticizing them," I told her.

I'd seen on the news that some people had been displaced altogether, had nowhere to go. At least I had my brother's place and Eugenia's place and about a dozen other friends and family members I could lean on. Still, what would I do if I'd lost that house?

We drove back toward Jason's house in silence, looking left and right at the things washing along the sides of the road. I thought of the towheaded boy, and the horse, and the man who had yelled "Thank you" to my retreating back.

We made it back to the interstate, and about halfway to Jason's place I saw a group of three people. A stout man was walking on the shoulder of the road with a small boy beside him. They wore backpacks. A lady with wild, uncombed hair walked right behind them. Eugenia stepped on the gas, and as we roared past, I looked over my shoulder after them, but the truck's tires kicked up water and blurred my vision. Hopefully, the trio had somewhere to go. A warm house with a hearth fire blazing. Hot coffee. A big, green pasture full of hay, and, of course, a horse barn.

Bernard Clay

elegy for an eastern kentucky grocery

how does an oasis drown in a desert
does it really matter once it's gone
once the aisles and everything in them
break down and compile
in a disintegrated heap
out on a parking lot in front of that
"hometown proud" sign taunting
those driving by
having to go forty more minutes
for some damn fresh food

over fifty years gwen's been here
where most black folks call sundown country
holding her hamlet down as a cornerstone
with reasonably priced meats and staples
right between cap's barbershop and super dollar
unapologetically black and appalachian
independent and owner
untouched all this time
'cause who's gonna bite
the only hand that feeds them
no matter the color

Bernard Clay

but why resurrect this place
it can easily be taken away again
that rockhouse creek isn't going anywhere
but that sentiment passes quick
'cause gwen's not built like that
she puts this community on her back
converting the skeleton of her storefront
into a triage and calls all in need
down from them hills
which in this age
attracts the viral content spotlight
leading to national morning show invites
to broadway tickets
to state acclamations from the governor
to comically and monetarily large checks
and corporate support to not just rebuild
but modernize the whole building
and of course gwen calmly
takes it all in with poise 'cause one day
that red iga sign will be a lit beacon
in isom again

Randi Ward

Dark Waters

Here
it's known
as
the devil's
piss—
damned
if it's
C8,
damned
if it's
C6.[1]

Note

1. *Dark Waters*, released in late 2019, is a movie about the real-life legal battle to hold DuPont accountable for contaminating the water districts surrounding its Washington Works facility in Wood County, West Virginia. DuPont knew for decades that perfluorooctanoic acid (C8), used in the manufacture of Teflon products, was extremely toxic yet neglected to inform the Environmental Protection Agency (EPA), its employees, or local residents. DuPont agreed to phase out the use of C8 in production by 2015 but replaced C8 with perfluorohexanoic acid (C6) and similar variants (GenX) without being required to demonstrate that these are any safer than C8. In April 2023, the EPA ordered the Chemours Company, a spin-off of DuPont that now owns the facility, "to take corrective measures to address pollution from per- and polyfluoroalkyl substances (PFAS) in stormwater and effluent discharges from the Washington Works facility near

Parkersburg. The order on consent also directs Chemours to characterize the extent of PFAS contamination from discharges." This marked the first EPA Clean Water Act enforcement action ever taken to hold polluters accountable for discharging PFAS into the environment. EPA, "EPA Takes First-Ever Federal Clean Water Act Enforcement Action to Address PFAS Discharges at Washington Works Facility near Parkersburg, W. Va.," press release, April 26, 2023, https://www.epa.gov/newsreleases/epa-takes-first-ever-federal-clean-water-act-enforcement-action-address-pfas.

Signs on the Jethro Amburgey Bridge along KY 160. © Melissa Helton

Marc Harshman
Headlines and History

> *But of that day and hour no one knows, not even the angels of heaven.*
> —Matthew 24:36

So not news, not exactly, but new enough,
    this suffering
    of the suffering.
And like the needle in the haystack
    the angels will be hard to find
    unless you count
    neighbor and kin.

And the rains come and the mountains shift
    enough, enough
    to show they're watching.

Listen.
There's nothing like the roar of water with
    no place to go.
There's nothing like the roar of water
    when you yourself have
    no place to go.

Watch that green field slowly become
    a brown smear almost innocent.

Marc Harshman

Watch the buildings jockey for position
   in the outrush to destruction.
Watch the skies empty their bowels
   hour after hour after hour.

What of the history?
Where's that in the headlines?
What of those clear-cuts, the strip mines,
   the fracking roads, the MTR[1]?

A lot of people are angry. Even more weep.

I want to hear the preachers preach justice,
   I want to see tears and anger turn
     to justice.

I want to hear the politicians . . .
   put up or shut up, as Father liked to say,
   I want to see their hands in the mud, see
   a light bulb finally flash when the history
   comes clear, their cooperation exposed
   for what it is.

What's happening here is happening elsewhere.
This is not poetry.
This is not even news.
This is you. This is me.
This is a life getting turned upside down.
These are buildings and bridges, homes and businesses
   wet through, molding, rotting, lost.
These are precious books, scores, banjos and fiddles ruined.
These are lessons delayed, bills continuing, Christmases postponed.
These are lives lost.
This is someone drowned, pulled under, gone, gone . . . gone.
This is not forgettable, forgivable, not easily explained
   as if just another headline, another banner
   streaming and running
   below the Weather Channel's pretty maps.

TROUBLESOME RISING

This is life broken, cascading under a broken bridge and
    looking to all the world as if it's just trash, junk
    when it was once all and more of a life, a life, lives
    lived under a more benevolent sky, lives lived
    in hope that justice might begin to stitch
    these mountains back together, might begin
    to ascend the statehouse steps and march
    back down with something more than
    meager compensation, false hope, lies
    that it won't happen again, won't become
    just another headline, more forgotten history.

Note

1. MTR refers to mountaintop removal mining, a means of coal extraction that permanently removes the tops of mountains.

Doug Van Gundy
The Flooded Town

Forget for a moment what's to come—reek of diesel,
swollen drywall crumbling to dust, liberated

spores of mold in full, black bloom—and notice this:
an empty town at the exact moment the waters crest,

nearly silent and everything still. Places that were
backyards and ravines and parking lots and will be again

are, for the moment, backwaters where ducks
and wading birds survey their temporary territory.

A green heron perches on a state road sign. A pair of mallards
paddle past the drowned steps of the Baptist church.

In the stillness, clouds are breaking up and sunlight
is spilling through. This is not the optimistic ending,

the silver sentimental lining. No one will be spared
their share of sorrow, their little ladleful of loss:

misery has a long reach. But for now, one might be forgiven
for seeing them as beautiful: these sunken, unpeopled streets.

Leah Hampton
Absence and Elements, a Prayer

IN HER BOOK ON THE FLOODING OF TEXAS AFTER HURRICANE Harvey, author Lacy Johnson asks two questions about that disaster: What do floodwaters obscure? What do they reveal? The result is a unique atlas, entitled More City Than Water, in which Johnson uses maps and diverse eyewitness narratives to find the disturbing answer: poorer neighborhoods and communities of color suffered worst from Harvey, largely because they were forced to. Not only did Texas turn a collective blind eye to climate change but Houston also willfully ignored infrastructure in marginalized areas for decades, and after the hurricane, FEMA only rebuilt the city's "better" neighborhoods. Texas knew the floods would come, and thanks to its many systemic inequities, the water damage was destined, even designed, to target its most vulnerable residents.

The lesson of Harvey is one we have already learned countless times in Appalachia, of course: when it comes to environmental or social injustice, our bitter past is the rest of the country's future.

I was reading Johnson's atlas when the Kentucky floods came. I was reading her book in Idaho, far from home, far from the rushing waters. I was reading not to understand floods or injustice—I already know too much about both those things. I was reading to find a way to write about fire.

It's not easy to be an Appalachian writer. There aren't a lot of jobs for me at home, so I travel a lot, taking gigs that force me to be away from my native western North Carolina for long stretches. My current gig is at the University of Idaho, where I help document wildfires in the rural

Pacific Northwest. I know a lot about rurality and climate inequity, what it's like to be poor and forgotten in a remote place, but I am often puzzled that Idaho wanted me for this job. What does a mountain girl know about a million-acre fire? I'm from the Blue Ridge, the soggiest place in the world.

My permanent home, a cabin in the woods west of Asheville, has no central heat or air. In summer, my spouse and I live with our windows open, and the house fills with water. We fight mold and gutter runoff. The house drinks everything in, sucking moisture from the earth and air. During late spring, the doors in the house start to swell, and soon they're so waterlogged they won't latch. By July, when the flood danger peaks, all our furniture feels damp to the touch.

There is always water. I have known much loss here, all kinds of pain, but our well never runs dry. My family comes from eastern Kentucky, where our losses are equally staggering. Somehow I always visit Kentucky when it's humid or raining, and every time I am struck once again by Appalachia's resilience, by our ability, even in the darkest times, to live without thirst.

Meanwhile, in northern Idaho, the world burns. I was a soggy woman stranded in the Northwest when the flood hit home. I was parched and scorched and smelling smoke, reading Lacy Johnson's atlas and trying to write fire stories. Then, early one morning, my phone began to whir with notifications: Kentucky, where seven generations of my family have lived and are buried, was deluged. Everything around me in Idaho was so dry it could burn in seconds. But at home, the place my body knows, on the land that lives inside me, there was too much water.

When you leave Appalachia, you must remake yourself and live differently on the earth. I rarely read or hear about this aspect of out-migration. We talk about leaving a lot, about missing home, missing the critters and smells of our childhoods, but we don't talk about how working-class Appalachians, who are born into a unique symbiosis with the land, relate to strange locations. For me, the most difficult adjustment of working temporarily in Idaho is not the homesickness or the infuriating lack of spice in the food. The hardest part is the way being there jumbles the elements. There is so much more *earth* in Idaho and not enough trees or water to cover it. When I am away from home, I long for emerald, deciduous canopy and find only scattered evergreens. The inland Northwest can be a sparse, denuded world, bedraggled and vast. Smoke hazes the air, and the wind carries strange birds, odd scents. Sometimes I think I would sell my soul to hear a babbling creek talk to me, to have a patch of dark moss pull me down onto its cool, moist pillow. Of course there's a lot of water out West, sure, but it keeps to itself. Huge lakes lie still and cold. The rain doesn't speak there, and the rivers won't let you touch them.

And from June to October, fire reigns. The air quality can be toxic, and sometimes smoke darkens the skies for hundreds of miles. I never know if I'll be able to breathe when I step outside; anything could catch light. I am accustomed to mountains being close and covered in mist or rainy puff clouds. In Idaho, the mountains are distant and unreachable, and any clouds—at least in the summer—are likely plumes of ash. Nothing about that landscape, lovely as it is, reminds me of water.

There is dread in living with wildfire, but also a weird kind of justice. If you're not from the Northwest, you quickly realize fire is only a "problem" there because most locals approach it from a white settler's perspective. The modern history of wildfire in Idaho, Oregon, and Washington is inextricably tied to the mismanagement and theft of land from Indigenous people and tribes. This region is *supposed* to burn, at least now and then. But white people like me keep trying to stop that from happening, so the fires, when they inevitably come, are more destructive than they would be if we would just leave the land alone. In the last hundred years, wildfire has become the only thing stronger than capitalism in the Northwest, the only force that fights colonizers with any success. It blackens suburbs in minutes, chars "investment" properties. But as with the flooding back home, very often the worst victims are the folks with the least privilege.

And yet I find myself admiring its power, almost rooting for fire, despite knowing that it can destroy me or the house of a friend or stop my breath. Fire belongs in the Northwest. It is part of that place. It cleanses and renews, gives plants fresh space to grow. And it forces me to admit my own complicity in the ravages of climate change. Fire hovers and waits, in much the same way water abides in every part of Appalachia. The very thing that most enriches a region can also pose the greatest threat. We are often destroyed by that which we know best.

To see fire and water as enemies is wrongheaded, I think. Why make adversaries of the elements? Fire and water are our neighbors, essential to every ecosystem, and they know us better than we know ourselves. We lose out to the elements sometimes, and other times they lose out to us, to our dams and developments. Either way, the land is not at fault when it burns or drowns us. Ultimately, our misuse of the land and our absence of a social safety net are to blame.

So when news reached me in Idaho of the tragedy that July, once again I had to rethink how I live upon the earth. I was not in your flood that day, but I know how the water behaved, how fast it came. I have been through my share of floods, and I felt sickened by every story, every picture. I felt the pain it caused you. I know what the flood obscured and what it revealed.

More importantly, I know that our regressive social policies and purposeful abuse of our environment are the root cause of the damage and that we will continue to reinjure victims long after the waters recede. Thousands are still displaced, uncertain how to continue to live in this perfect, impossible place. What saddens me most is we are doing that part to ourselves.

As a reluctant and only part-time out-migrant, I felt a surprising clarity after the Kentucky floods. Studying Northwest fires helped me realize something about my relationship to Appalachia. Because the thing is, the University of Idaho *did* invite me, unlike the snooty institutions of Asheville, where despite my impressive résumé, I rarely have any luck finding work or support for my writing. I lack the generational wealth and Biltmore/Florida pedigree necessary to thrive here. My in-your-face politics ask too much of my neoliberal peers. And my MAGA-hat neighbors, who refuse to believe in climate change or higher education, view me as a threat. I talk too much about poverty, about elitism in academia, about true, hard climate justice. In short, I make Asheville uncomfortable with itself. At home I am undervalued, exploited, washed up. And elsewhere in the region, especially Kentucky, there is little opportunity for me. But *Idaho*, of all places, found value in me that my home cannot or will not.

The Northwest is different but no less beautifully dichotomous than home. And it summons me in a new way, asks things of me that Appalachia never has: What if you had space to grow after loss? What if there was less water, more fire? Could you have a new conversation with the land? How would it feel to be *invited in* to a place and its climate suffering instead of just born to it? How would it feel to be a welcome stranger? To live, for once, in hope?

As I write this, I am sitting in my little cabin, where my husband and I are wrestling with a decision many readers will understand—stay or go? I've recently been offered a permanent job in Idaho. We can make a new start, or we can remain in Appalachia and keep struggling to get by. Face the coming storms in this run-down mountain house I adore, with birds and trees I cannot imagine life without. Or jump into a fire. Stay and fight or leave and thrive. My options in Idaho are largely the result of my white, educated privilege, but this same question surely haunts many Kentucky flood victims, especially those from underserved communities. The place we love has hurt us, and the hurt keeps right on coming. Home does not seem to want us anymore, and I do not know what to do about it.

For the first time, after always looking down my nose at "expatalachians," I now find myself tempted to leave. Like many great hillbillies in history, I am drawn to the possibility of the West, the opportunities it offers that I have been denied at home. I am also aware that I would not be missed. The birds and trees, the babbling creeks and dark moss, would be fine without me. Perhaps they'd even be better off. Perhaps I would too.

Perhaps my heart will break if I leave, and it will never heal.

Or perhaps I should leave the land alone, so she can heal herself.

Whether I stay or go, I lose. But loss cleanses, instructs. Floodwaters wash away the past; fire returns it to ash. Tragedy readies us for change and regrowth, and Appalachia needs change. These floods, these constant injustices—they are the land teaching us, telling us we have got to do better.

Kentucky friends and ancestors, I do not know if I can remain in these hills. I wish it were easier to stay with you, to make Appalachia whole again. I wish you loved me, and yourselves, enough to keep more of us around.

As you recover, I can only offer this prayer: May we learn from what has been revealed by these floods. May we change for the better. May we give ourselves permission to stay or go without bitterness, with the understanding that it is not the elements who hurt us but ourselves. May we hear a message of love from the land that cries out for healing.

May fire and water do as they will. May we respect their power.
May we now live differently on the earth.
And may we find hope, somewhere. *Amen.*

A young volunteer opens the door to Uncle Sol's Cabin. © Melissa Helton

Pauletta Hansel
No Friends of Coal

What a friend we have in Jesus, / All our sins and griefs to bear!
—Traditional hymn.

And he shall break down the house, its stones and timber and all the plaster of the house, and he shall carry them out of the city to an unclean place.
—Leviticus 14:45

And when the unclean place
is the silt pond
up by the strip mine and the gray
of its waters broke free
from its rickety dam
and carried away your house built
wise as you could, what else but a lawsuit
to carry the blame
back up the shorn mountain
where it belongs?

   "What a friend we have in . . ."

Jesus wept
and I'm willing to wager he would again
at Blackhawk Coal and their claims
of "our people" and "deeply impacted,"

Pauletta Hansel

their sympathies and their support.
And the mud had not dried
when they posted their notice
"Intent to Blast"
over where the door stoop had been
before the flood.

I know there was rain,
too much, too fast,
another one of those thousand-year floods,
but what kind of flood
carries its little ashy silt pond fishes
down from what's left of the mountain,
not up from the creek?

> "Within minutes of the color change, the water rose so high
> that it picked up homes, cars, sheds, boulders, trees,
> staircases, swing sets and swimming pools."

"No Friends of Coal"
reads the headline about the lawsuit
in the local weekly paper
underneath the photo of what used to be
that young man's house,
now torn down to its timbers,
its moldy plaster spread out on the stony ground,
the raptors circling just beyond the frame.

> "And the Philistine said to David, Come to me,
> and I will give thy flesh unto the fowls of the air,
> and to the beasts of the field." 1 Samuel 17:44

That young man's name is not David.
I wish he were David, delivered
"out of the paw of the lion,
and out of the paw of the bear,"
and "out of the hand of this Philistine."

TROUBLESOME RISING

So, yes, I'm feeling biblical,
Old and New,
calling on Jesus
to back up his people,
once clothed and fed by coal.
The ones who lost everything
there on Lost Creek, along River Caney.
The ones that are fixing to lose
every friend they thought they had.

   "What a friend we have in . . ."

Brother, I am here to tell you,
no matter the money they used to pay,
you don't have a friend in coal.

Christopher McCurry
Brothers 12 & 14

Chained in her pen
our dog dies in the flood.

You cry because you love her
and she loved you back.

I don't cry because I hate
the stake, the chain, the dog.

Each one a tether.

Hundreds of water-blackened
leaves hang on the fence.

Their stripped veins
worshipful hands

reaching for the world
jerked away. We hear

on the news that rescue
crews couldn't find

TROUBLESOME RISING

the mother or the child
swept to where we all

eventually follow.

Some bent low
like the grass,

others strung out
like a dead dog.

Frank X Walker
Elvis

captured Gwen Christon's story
in a song
almost fifty years ago,
except her lost love
was a grocery store, in Isom.

And the long walk
was more of a swim.

He was right about the dozen towns.
He was right about the lonely back roads.

He must have seen Letcher County
and its surrounding hollers
under *six feet* of muddy water.

He must have had visions
of school buses only visible
by their roofs.

When he said,
love too strong to let you go,
he must have known about
the decades she would pour

TROUBLESOME RISING

into her IGA,
the people she would feed,
the community she would nourish,
and her own flood of tears
when the store was gone.

He was right about the old men
and the general store.

He was right
about the preacher man
and his prayers,

but he was mostly right
about *the cold Kentucky rain*.

Nikki Giovanni

Where Was the Music

At first I thought
Of baths
A warm bath my grandmother
Would run
For me
Then rub my back
But No
This was angry
Running away from smiles
The water swept off
My feet
Pushed madly
Into the house
And I grabbed my grandmother's
Bible and my little dog
And climbed to the roof
To wait
For relief

I knew there was a moon
Up there
But I just couldn't hear
It sing

Lee Smith
River Rising

I WAS IN MAINE WHEN IT HAPPENED, WHEN I FIRST HEARD ABOUT it—the horrific flood that tore through eastern Kentucky, devastating the Hindman Settlement School, my own spiritual "writing home," just as surely as floods had devastated my own parents' home and my father's beloved Ben Franklin dime store again and again in my hometown of Grundy, Virginia.

My father, Ernest Smith, loved Grundy and wouldn't ever leave it for anything, not even after my mother died and we tried to get him to move to North Carolina with us.

"No, honey," he always said. "I need me a mountain to rest my eyes against."

The mountains, I always believed, determined everything.

But now, looking back, I'm thinking that maybe the Levisa River behind our house was even more important. It's kind of a shock to realize this. But I guess it makes sense. Our mountains in Buchanan County are just like the mountains in eastern Kentucky—pretty much straight up and down, and many of them have been strip-mined, which assures a fast runoff. So when the creek has to rise, there's no place for it to go.

My own little universe, my childhood home and my father's Ben Franklin dime store, were both located right on the banks of the Levisa River, a pretty rippling stream on most days—except when it turned black from the coal they were washing upriver. Then no amount of laundering could get your clothes clean—and since most of mine were handmade by my mother, a home economics teacher, this was a catastrophe.

But I loved the river behind our house and the Norfolk and Western trains that roared along the track beyond it several times a day, carrying coal. On the edge of the riverbank sat the little writing house that my daddy had built for me—and then had to build again after every flood. Sometimes it was made from a wooden crate that had been used to ship merchandise to the store—or a prefabricated toolhouse or little shed they sold in the basement.

How I loved to write in my writing house and play down by the river, where I had many imaginary friends in addition to my best friend, Martha Sue Owens, and the other neighborhood kids. The river was idyllic then—paradise! It's hard to realize it could become a raging torrent as it did in the great flood of 1957, which literally wrecked the entire downtown of Grundy, including the dime store.

I will never forget the huge catfish I found flopping down the dime-store stairs into the basement level, the water-filled toy and gardening section. It used to be my job to take care of the little turtles they always had for sale down there, but now they were dead and floating everywhere, with those little roses painted on their shells. It was horrific.

I came back home from North Carolina to help when the Levisa flooded in 1977. That flood killed three people, devastated 90 percent of the downtown businesses, and caused $100 million in damage throughout Buchanan County—the equivalent of $500 million today. My parents' home was also ruined, with the muddy water rising above the countertops in Mama's beloved kitchen. I remember gathering up floating pieces of the parquet floor in the dining room to keep as building blocks for my little boys back in North Carolina.

A nervous woman even in the best of times, my mother never really got over it.

Daddy didn't either. Grundy had had nine major floods since 1929. I remember so well how he never slept when it rained. All night long, he was walking back and forth with his flashlight to "check the river" again and again and again. He had a huge steel flood door constructed at the back of his store, which was put into place every time the river started rising.

In 1992, at eighty-two, Daddy finally closed his beloved dime store "due to lack of business," as he said. It was horribly fitting that he died on the last day of his going-out-of-business sale—though if any of my creative writing students had written that in a story, I would have said "too pat"—or "schmaltzy."

The dime store and my parents' home—along with a score of other homes along the river plus the entire former main street downtown—are long gone now, but because of the drastic and daring Grundy Flood Control and Redevelopment Project, a historic collaboration among the US Army Corps of Engineers, the Virginia State Department of Transportation, and the town itself, Grundy has risen like the phoenix from the rubble, miraculously holding its own during the recent flooding.

Thinking back now, I understand the terrible toll this constant flooding took on both of my parents, a major factor in the serious depressions they both endured, hospitalized in several different mental hospitals, sometimes for months at a time. This was a major factor of my youth.

I understand now that the Levisa River and repeated flooding determined my parents' lives—and life in Grundy as a whole—far more than the mountains themselves did.

The cleanup of flooded spaces begins. © Tyler Barrett

Annie Woodford
Five-Hundred-Year Rains

Little River Canyon, Alabama: 2013

    When we walked,
we walked through water.
In dripping woods,
Bellwort and Bloodroot
    pushed through duff.
All trunks and rocks
    were wrapped in moss.
The leaves rang.
The devil beat his wife.
Fattened Resurrection ferns
    filled the forks of oaks
with even deeper green.
    We took shelter
under a rock pavilion—
sandstone cut from the quarry
less than a mile away,
    the artistry of CCC boys,
now old men or dead,
    evident in the extra arches,
the chimney big enough
for my six-year-old

TROUBLESOME RISING

to stand in, a dreamer's
   vision of a fireplace
(*and in the castle, the king*
   nodded by the fire).
Every trail we walked
was stream. Water
gushed from cracked
   rocks, full to bursting
for the first time
   in local memory.
That late April
was the beginning
of many floods. The pop
   of rain on a tarp.
The shifting of weather
   systems thrown off-kilter
by heat. We went
to the gorge, the giant
falls. Did de Soto's
   scouts sense a shift
in the landscape?
   The arrival of awesome
geological fact? Birds flew
below us at the overlook.
The falls were so loud
   we could not talk.
I started to learn
   how water can change.

Tina Parker
Learning from Home

While places like ours wear on the nation's nerves
We hit the hairpin curves.
We show our daughters strip-
Mined mountains, railbeds, boarded-
Up UMWA[1] buildings.
We want them to learn from what we've lost.
Want the Confederate Flag
Curtains and car decals to take
On greater meaning.

Oh how they long to vote.
They want to leave,
Go anywhere big.
They've learned what it is to hate.

Look at the pretty birds they say
And wave middle fingers out the car windows.
Even in their dreams they cuss
And rail *No effing way.*

Note

1. The United Mine Workers of America (UMWA or UMW) is a labor union best known for representing coal miners.

Savannah Sipple
Ain't No Grave: A Story

ALL IT TOOK TO MAKE ANNA DECIDE TO LEAVE THE HOLLER WAS A body slam. At church, Mrs. Fugate got filled with the holy spirit and knocked Anna on her ass. About a year ago, Evangelistic Tabernacle had started filming their services so they could be aired on the internet, and Anna was operating one of the cameras. She was a videographer and video editor, and it was her job to help make the gospel look and sound like Jesus was in the building. The church band started playing "Ain't No Grave" as a way to get people out of their seats. It was a crowd favorite both in the church and with the viewers. They started in on the chorus for a third time, and their tempo and volume increased as they sang about graves and trumpet sounds and holding the body down. Mrs. Fugate was shaking and shouting, almost like she was about to gag, except she held her hands up in praise and hollered, "Woo! Woo!" Then she started speaking in tongues and took off running around the church. Just as the preacher was encouraging people to give themselves over to the Lord, to let his power in, Mrs. Fugate gave a great shout and ran right over Anna. If her tackling Anna and the damn camera wasn't enough to show the audience what the power of the Lord can compel us to do, what would be?

Anna lived on the edge of Bear Creek, nestled at the base of a mountain whose top had long been blasted flat. She came back home after college and moved in with Granny to help take care of her. Between Granny, work, and the church, Anna hadn't had time for much for several years. She worked at the local high school teaching English, and the church paid her a little for the video editing. It was just her now. Granny had given up the ghost a little over a year ago, and even though Anna always said she'd leave the holler for good when Granny passed, she stayed on in the trailer. She kept the garden going and delivered food to the older neighbors, friends of Granny's who'd lived in the holler for as long as she could remember. Anna wasn't sure why she hadn't left, especially since it was the thing she'd planned for since coming back home seven years ago.

Anna threw herself into church work to keep herself busy, even while Granny was still alive. She worked with the teen youth group, volunteered to help clean the church, and even helped with some of the regular maintenance. Granny led a prayer circle that met once a week. Anna often cooked or baked something for the ladies to share, and if she wasn't teaching, she joined them. All that praying and serving the Lord, and none of it changed the fact that Anna was a big ole queer. Her service to the church, Granny's health, and her lack of any dating life were just enough for folks to look the other way, but they all wondered about her. It wasn't how she dressed that gave her away. Plenty of mountain women spent most of their time in Carhartt and overalls. And since the church was lax in its dress code, many of the ladies wore slacks. The thing that rested uneasy in other people's minds was her lack of a feller. Sure, other girls stayed single as they approached thirty, but Anna had never brought a boy home or even tried to find some unsuspecting young man to fake a crush on. She did not talk about love or affection, and if anyone dared to ask when she planned to find a husband, she had to catch herself so she wouldn't say, "Never."

But that all changed with Granny gone.

Anna's friend Jake, from college, was queer as a three-dollar bill, and he'd been trying for years to get her to move to Lexington and out of the mountains. Once in a blue moon, she'd drive up for the day to meet Jake. They'd start the day at a coffee shop, where they'd catch up, and then they'd make the rounds of the bookstores in town before ending the day with dinner somewhere. She often tried to talk him into sushi or Thai—something she couldn't find back home. When Granny died, Jake drove all the way from Lexington in the middle of winter and stayed by her side during the visitation and the funeral, and he didn't leave until the last cousin from out of town was long gone. He made her promise to come for a proper visit, a multiday visit, as soon as she could. Jake knew

Anna was gay long before she wanted to admit it, and he'd been dying to get her out of the holler and into any place that had other queers.

Anna tried to back out. First, she made an excuse about bad weather and the power going out and the neighbors needing help, but then spring came. She kept talking about the garden and how much work it was to get it ready, until finally Jake said, "Run the damn plow and get your ass down here." So she went.

Her first night in town was easy. Jake invited a few friends over, and he and Anna cooked dinner together. She'd brought canned mustard greens and corn from home. Anna fried the corn and cooked the greens with a little piece of middling meat. They served it with Jake's roasted ribs and homemade bread. His friends ate enough to founder as they sat around the table, drinks in hand, laughing and telling tales. They all drank wine, but Anna sipped a bourbon. They insisted that Anna come out with them the next night, a Saturday, to someplace called The Bar Complex. Jake had talked about it before, but she'd never been and wasn't sure what made it a complex. His friends held their wineglasses in one hand and talked about the fabulous queens she'd see there. A drag show. She'd never been to one of those either.

"Come on, honey, it'll be a blast," Jake's friend, Connor, kept telling her. "All those gays and lezzies in one big place. It's like a family reunion."

"It's good to be among your own people," another said, patting her shoulder. Anna was out to herself and to Jake, and she didn't mind that his friends assumed or maybe even knew she was a lesbian, but she had never really been somewhere to meet other gay people. A friend of a friend was one thing—low-key, low stakes. To be surrounded by queers was another. It was both a thing she longed for and a thing that scared the shit out of her.

Jake had to work the next day, so Anna spent the day reading in the armchairs at her favorite coffee shop. She'd pulled a few books from Jake's shelves to borrow, queer books by queer authors. Every time she was in town, he'd lent her a few or recommended a couple at the bookstore, and she had somehow managed to keep them hidden from Granny—in plain sight, really. Double-shelved with her books in her bedroom.

That night, she showered for a second time that day and tried to get ready for The Bar Complex. She knew it was a drag show, and she'd tried to figure out how to dress by looking at pictures on The Bar's social media pages, but the truth was, she wasn't sure how to look like a lesbian. Maybe the stereotype about lesbians being so butch wasn't true. Besides, Jake had told her she could not show up in her work boots, so her assumption was that she needed to "girl it up a little" and go a bit fancier for the night out. She'd brought her nice church pants, black slacks, and a red top with a ballet neck that in all reality made her feel a little uncomfortable because it dipped so low. She tried to put on makeup but didn't get much further than mascara. The rest of it felt too heavy, too hot on her face. She wore black flip-flops, Crocs, because they supported her worn-out feet almost as well as her boots. When she came out to leave, Jake looked her up and down and smiled. "You look nice."

"And you are full of shit."

"No, no! You do! You look good." Jake tried to hide the pained expression on his face.

"What am I supposed to wear?"

"You look fine. Come on. Let's just go and have a good time."

Anna sighed, grabbed her wallet, and followed Jake out the door.

When they arrived at The Bar—no one who went there included *Complex* in the name, Jake had explained—they were greeted at the door by a short stout woman with cropped silver hair. She checked their IDs, collected their ten dollars, and stamped their hands. Anna followed Jake into the club, past the first bar, which was swamped with men dressed in tight, tight pants and fitted shirts, with perfectly groomed hair. Several of them said hi to Jake.

They made their way to a curtained doorway, and Jake turned to make sure Anna was still behind him.

"This," he said, "is the best part."

Through the curtain was a large room with a stage. Tables and chairs circled the stage on three sides, and long booths ran along the side wall. Tucked in one corner, to the right of the curtained doorway, was a smaller bar, but it was even busier than the first. Jake went straight to it and wiggled his way in to make an order. Anna tried to follow, but she kept getting distracted. She'd never seen so many gay people in a room at once. There were men and women coupled up, grouped up, holding hands, sneaking kisses, and wrapping their arms around each other. There was one group of what appeared to be straight women out for a bachelorette party, but most everyone else seemed to be queer. Some had piercings or tattoos, some were scantily clad, and others looked to be dressed almost . . . normally. Was that the right word?

Anna felt so out of place. She focused particularly on the women, their low-cut shirts, their tight jeans. And the butch ones, they had on jeans and button-up shirts or plain T-shirts, and some even wore muscle shirts. Anna looked like she'd tried to dress straight and failed. Not quite queer, not quite right. Before she could panic too much, Jake pulled her to the bar to place an order.

"What'll it be, honey?" the bartender asked. He was as short as the doorwoman and looked to be even older. He wore black pants, a white shirt, and a leather vest. His accent sounded like home. Anna loved him immediately.

"What do you recommend?"

"How about a Come Fuck Me?"

"What's that?"

"It's our signature drink. You'll love it, honey."

The man mixed what appeared to be several types of alcohol with some pineapple juice and grenadine, threw a couple of cherries on top and a straw in the cup, and set it in front of her.

"Baby, I just popped your cherry. Welcome to The Bar. That'll be ten dollars." She grinned and paid.

Jake led her to a table that was to the right of center stage—perfect for watching most of the room and close enough they could easily tip the queens. His friends from dinner the night before soon found them, and Anna started to feel more at ease. It might have been because she was on her third Come Fuck Me. There was something beautiful about the way people were coming and going, wiggling their bodies to the music, cheering the queens. Folks stood at the edge of the stage with one-dollar bills in their hands, waiting for the queen to come by and grab the money. Some of them were blessed with a kiss on the cheek. Some of them blushed. Anna tried to take it all in, to breathe in the knowledge that all these misfits belonged here, together, when the next queen caught her eye.

She was tall and tan with a neckline that plunged damn near her belly button, the curves of her breasts held in by the tight edges of the dress. She started singing along to some song Anna knew but couldn't fully register because she was so taken in. She was turned on by her curves, her face, her entire body, and the way she danced to the music. Anna's ears started ringing, and in the back of her mind, she could hear the lyrics to the church band's signature song. Over and over it played about the band of angels "comin' after me." Why in the hell was that song stuck in her head now, of all times?

Jake elbowed her. "Hey, Anna. Anna!"

Anna peeled her eyes away from the queen. The music faded.

"Go tip her."

"What?"

"Tip. Her. Put your tongue in your mouth and go give her some ones."

Anna swallowed hard, downed the rest of her drink, and stood. She felt lightheaded but made her way to the front of the stage. She held her hand out, with two ones tucked between her fingers. The queen shimmied her way to Anna and stood in front of her, lip-synching. Anna handed her the ones, then pulled two more from her pocket. The queen plucked them from her fingers, and Anna got two more. The queen leaned down, putting her breasts right in front of Anna's eyes.

"Here ya go, honey. Slip 'em in there."

Anna tucked the ones underneath the top edge of the plunging dress, and the queen wiggled her chest and gave Anna a peck on the cheek.

"Thank you, baby." And she sauntered away.

Anna blushed bright red all the way to her ears as she floated back to the table. The rest of the night passed in a blur of dancing queens, laughter, and people watching. Hours ago when she first arrived, she had felt like this was absolutely not the place for her. But as she watched all the people, all the queer people, happy and carefree, she thought, maybe, she could find a place to belong.

Anna went back and forth to Lexington as much as she could that spring. And over the summer, she wasn't teaching, so she had free time during the week. She realized she could catch either the drag show on Friday night or the one at 10:00 p.m. on Saturday and still be back in time for church service on Sunday. She met more queer people and even let Jake talk her into joining a dating app for women. She'd only been on one very awkward coffee date, where she couldn't figure out what she was supposed to do, so she acted like a man, opening doors and paying for everything and trying to be somewhat macho, but mostly she felt nervous. But still, it was a date. She had queer friends. She realized she wanted to move, and she'd been watching the job ads since the end of the last school year, but English teachers were a dime a dozen. Then the church decided to record their services.

They asked for her help, and the money they paid wasn't great, but it was extra money that she could set aside and save for when she did want to leave. This was the start of her plan. She began putting aside everything she could. She even accepted the few dollars the older neighbors offered when she took them food from the garden or helped with the odd job. All of it cut into the time she could spend in Lexington that fall and winter, but in the long run, Anna hoped it meant she could move there permanently, even if she ultimately had to give up her teaching job.

No one at home understood why she seemed so hell-bent on living in Lexington when she had a perfectly good home and job. There had been a few rumors flying around town, likely started when, at a drag show one weekend, Anna bumped into an old rival from her basketball days. Folks had said for years the woman was a lezzie, and seeing her at The Bar confirmed it, but Anna also knew that their shared otherness wouldn't protect her. The woman had always been spiteful and loved a good story, and by the time Anna walked into church on Sunday, folks were already whispering. The whispers kept coming over the next several weeks, to the point where the preacher had called her into his office one Saturday under the guise of seeing how the video work was going. But Anna knew.

"I guess you're wondering why I wanted to meet with you today."

"I guess so."

"Well, I wanted to see how things are going. Is the video editing getting to be too much? I know it's a lot to ask for graphics and text on the screen."

"Not really. Once I figured out how to add them, it's pretty easy from there."

"I see. And how's everything else? The garden?"

"It's good. I cut back a little this year since it's just me. I still have plenty to share, though."

"That's good; that's good. I know those old folks have come to rely on you, in a way, Anna. I hear you've been traveling some lately. Up to Lexington? You have friends there?"

"I do. A friend or two from college. It's good to get to see them more."

"I see. I'm sure. I hope they're good friends to you."

"They are. They've been a big support since Granny passed."

"And you're not getting into any trouble in Lexington?"

"Any what?"

"Any trouble? A couple folks have mentioned some concerns. And I'm not saying there's any truth to it. Who knows how these rumors get started. I'm only saying that as our videographer and editor, you represent the church. And we wouldn't want to have any issues come up."

"I see. Well, I'm not getting into any trouble."

"That's good. That's good to hear. You know, there are plenty of fellas around here who I'm sure would be happy to spend some time with you. You know, to get to know each other better. With your granny gone, it might be time to start thinking about settling down."

"I'm not really interested in that right now."

"I know there's plenty of time, but you might want to think about it. It takes time to date and find the right man. Plus, having someone here might keep trouble from finding you. The devil sneaks up on us when we're least expecting him."

"I'll keep that in mind. Is that all, sir?"

"It is. Thank you for coming by, and, Anna, if you need anything, my door's always open."

Anna left the preacher's office with a secret clutched tight to her chest. She'd found an apartment and paid the deposit. By the end of the summer, she'd be gone from this place. She'd lost too much. When she was in the garden or at the edge of the creek with a book, she felt like she could stay in the mountains forever. But sometimes it threatened to drown her, the pull and tug of loving a place and knowing she couldn't thrive there. Her books couldn't replace having true friends, and she could grow a garden anywhere. Her love for the place would never outweigh the sad looks she often got when she showed up alone or the rumors that were whispered in the church pews when people thought she wasn't within earshot. There was no doubt where her principal stood on the idea of Gay-Straight Alliances in school or even of allowing students or faculty to share their pronouns. And for all the preacher's talk of support, the message from the pulpit had grown hateful, full of assumptions and outright lies about "the gay agenda." It all felt like rocks in her pockets. Anna left the preacher's office and went home to start looking at what she should pack to take with her and what she should sell.

The next morning, when Mrs. Fugate plowed her over, Anna's entire body rattled. And the preacher implored her to get up and keep on filming the good Lord's work. She was liable to be bruised all to hell, and the only thing that man cared about anymore was the number of views he could get. Mrs. Fugate was already up and running again, and as Anna pulled herself off the ground, no one stopped their worship to help her. She stood, set the camera upright, and tried her damnedest not to cry.

At home that evening, she let the tears flow freely as she mixed up some corn bread to eat with the fresh-picked green beans and ripe tomato from the garden. She was sore, but grief overtook her. Granny had been the thing that felt like home, and that was long gone. Anna knew that leaving didn't mean she was turning her back on anyone or anything, but that didn't lessen the pain of letting go. She tried to focus on the future, on the life that could grow from nothing once she got to a place where she didn't feel so wrong all the time. Anna distracted herself by staying up late and packing some of her books. Sometime around midnight, the sound of rain on the roof lulled her to sleep.

Around 3:00 a.m., Anna was jerked awake by her phone ringing. It was Old Man Jenkins from up the road.

"Anna, girl, ain't you heard the water? The creek's up. You gotta get outta there."

Anna jumped from bed, threw on some clothes and her boots, and turned on the porch light. In the shadowy distance, she could see the creek swelling over the bank. It was already up in the yard. She ran inside, piled as much as she could into a backpack, threw some random things into her car—her granny's cast-iron skillet and favorite quilt—and took off toward the Jenkinses' house. They lived closer to the bridge and were on slightly higher ground.

Mr. and Mrs. Jenkins were waiting for her on the front porch.

"Come on, girl. Get what you can out of that car and let's go."

"You all ride with me. We gotta get out before the water washes over the bridge."

"Girl, that bridge is long gone. We're heading up the mountain on foot. Get your stuff and come on."

Mr. Jenkins led the way with a flashlight, and Anna was surprised at how well both of them were able to manage the wet terrain at their age. She felt herself slide in the muck more than she'd like to admit, but the two of them kept trudging uphill. Finally, they came to a flat spot where they could look down at what was happening. The water took Anna's car first because she'd been rushing too much to think about where she'd parked. The Jenkinses' chicken coop and front porch washed away, and by the time morning came, it looked like the rest of the house was at least three or four feet underwater. Anna knew this meant her trailer was gone. Her garden. All of her things. Could she leave if she didn't have anything to take with her? Could she really walk away from folks who might need her even more now? As she watched her world get washed out, the same old song kept running through her head. She couldn't think about what might have happened to the others in the holler. "Meet *them*, Jesus," she thought. Surely they, too, made it to higher ground.

Daughters of the American Revolution and writing community volunteers work on archive rescue in the Gathering Place. © Melissa Helton

Sonja Livingston

Reliquiae Diluvianae
(Relics of the Flood)

BLACK CAT FIGURINE WITH CHIPPED EAR. CHILD-SIZED LEATHER shoe. Wooden flute. Embossed ceramic cup and smattering of blue plates. Shuttle from a nearby mill. Hettie Ogle's keys. A hymnal. Two serving platters. Wooden tool chest found seventy miles west, tools intact. A sea of upturned tables, herd of three-legged chairs. Enough buttons to fill a washtub. Doorknobs, balls of half-grown cabbage, window frames. A wooden oar. Portraits tucked into the Thomas family Bible. Worn cameo brooch, featuring a woman, white as a ghost.

Framed map, counties blotted away. Silver spoon. Medicine bottle caked with mud. Jar of honey, still edible. Wicker chair, soft as flesh. Inkpot, green glass. A lady's tortoiseshell comb. Canning jar with peach residue. Carved decoy (mallard). Recipes for holiday cinnamon rolls and Aunt Margaret's molasses cookies (illegible). Daisy Heslop's underdress. A single gold tooth.

The monstrance (empty eyed, half-gilded) from Saint John's Roman Catholic Church. Handkerchief from C. T. Schubert's trousers. From Vincent Quinn's pocket: change purse, phrenology handbill, slender red notebook. Pitchers for butter, cream, and milk. Straight razor. Handbell. Tin binoculars. Coffee pot. Wooden oar. A photo of the Waters family on their front porch. John Burns's calico quilt, used as a rope to pull people from the surge.

Hettie Lininger's photo album. Minnie Huston's trunk (washed into a downtown church). A jacket (checked, satin trimmed) worn by nine-year-old Bertha Morgan while scaling debris to reach the hillside. Trousers

Archive photos of James Still, Jean Ritchie, and Verna Mae Slone drying in the Great Hall. © Melissa Helton

(striped, brown) worn by Chalmer Barley (five), the only child in his family to survive. Ceramic doll. Padlock. Clockface. Lock of hair. Dr. Wakefield's medical kit. The cracked head of an alabaster Saint Joseph (eyes downcast, tired). Bottle of floodwater, black. A thirty-nine-star flag, hung the day before (Decoration Day) when families festooned graves with posies and wreaths, never once guessing how soon they'd be with their beloveds again.

Note

The Great Flood of 1889, Johnstown, Pennsylvania. After the collapse of the South Fork Dam, Lake Conemaugh rushed fourteen miles upstream, and the sudden force of nearly four billion gallons flattened everything in its path. The debris-spiked water rushed westward, carrying livestock, houses, railroad cars, and miles of barbed wire. In the end, 2,208 people died, with bodies found as far away as Ohio. When the waters finally receded, the pile of debris was seventy feet high and thirty acres wide. A "flood relic" auction raised money for a relief fund. Many of the items listed above are on display at the Johnstown Area Heritage Association's Flood Museum. See https://www.jaha.org.

Scott Honeycutt
The Only Prayer

The only prayer I recall saying word by word was an appeal
uttered one Sunday morning while kneeling by the Kentucky River.
It was so early in spring that waterside privet leaves were colored
like young limes, and sycamore buds hung closed-off and lancelike.
My daughters, too young to appreciate how seasons won't wait for love,
laughed and broke off running through the open forest.
They were themselves prayers.

The words I lifted were too pagan, I suppose, for the Methodist
steeples that shot up along those piney back roads, those red clay mornings.
But the river god, with its brown, fluid arms, answered.
It heard my howl and delivered the only news needed in this life:
Each day is more feast than gift, and we, the consumed, have for now our one
chance to say, *Thank you, I accept.*

George Ella Lyon
Don't Tell Me

you were terrorized by Troublesome
which was only a creek doing what creeks do
when rain torrents ten inches in hours
onto already saturated ground.
                         So it's rain's fault then—
unnatural, unceasing, giving its all, till it's over
the steering wheel, under the door, filling your room.
It's tearing up trailers, hauling off houses, swilling
children from their parents' arms.
                         But no, it's not
rain any more than it's Troublesome. It's your
kind, blowing up mountains for coal. It's people
greedy for ease and power who'll
heat the seas till the whole planet boils.
                         Don't curse Troublesome,
don't blame rain. When the water goes down
find a sliver of mirror, clear it as best you can
without clean water, then have a look.
There's the face for your Unwanted
Poster. There's your force behind
the flood.

IV | There Is Nothing Untouched

Staff and volunteers unload donated supplies. © Melissa Helton

Lisa J. Parker
Surge

Phone calls drop one after the other,
my sister somewhere between Hindman
and Hazard, dying phones passed between
family and friends—
a relay,
an accounting,
a mapping out
of dry road,
of safe passage.
I yell into the phone
I'm coming for you!
over and over into the static line,
and drive south through an obscenity
of Virginia sun that beats the tops of my hands
on the steering wheel, the lift and fall
of mountain after mountain, until I reach her
in Tennessee, hold her until we can both
stop shaking.

It's weeks before I'm at Hindman,
my jeans covered in drying muck from recovery work
in Letcher County next door, I walk on the strange silt
by the still-standing bridge, to the tree
where a picnic table should be,

TROUBLESOME RISING

where Ron Houchin should be
lying on his back smoking a stogie,
his Ed Hardy hat pulled down over his eyes.
The rockers are gone from the porch where I first sat
with Verna Mae, laughed with Lee and Silas,
watched Mr. Still hold court.
Water and mud lines mark all the buildings,
the apartments where my friends stayed,
where they ran out just before doors were knocked
off hinges as the waters crashed through,
where thirty years ago I rested my chin
on the guitar I held in my lap, listened
to Betty Smith singing "Barbry Allen"
and Cari Norris wailing "The Two Sisters."

Heat bugs call and I stand still
in the breeze blowing down off the hill,
down from Stucky where so many sheltered,
where the newly homeless still find respite,
find water from the well brought back
to full function by Moses who has been there
for as long as I can remember.

I sit by a rose of Sharon far enough up the hill
that it wasn't plucked away,
close my eyes and hear the risen creek,
its constant charge against rocks
and limbs caught in roil and eddy
that snap like bones at the bank's edge, listen
to occasional birdsong from robins and wrens
who keep to the high branches of poplars and oaks,
above limbs holding screen doors and sofas,
their wary calls carry down on the breeze,
the treetops heavy with their watching.
There is nothing untouched.

Doug Van Gundy
The Morning After

On the day after the flood, I baked a pie.

I wanted sweetness. I hungered for it.
I wanted the house to smell of cinnamon and butter
    and the resonant souls of apples. I wanted
    steam to fog the windows and peelings in the sink.

I didn't know what I was doing, but I kept going.
I called out to family and the internet for guidance.
I learned there is no one way to bake a pie,
    no one way to eat it.

I measured flour and weighed out Crisco,
    added a pinch of salt and a splash
    of vinegar as my matriarchs taught me.
I sliced apples parchment thin and macerated them
    in sugar and spices and the juice of a lemon.
I tossed in a handful of raisins to plumpen and burst.
I blind-baked the crust, then filled it and baked it again.

TROUBLESOME RISING

I waited for the alchemy. Tomorrow, there will be plenty
    of time for sorrow, plenty of time for shoveling
      out basements and sharing sympathy and patting hands.

Today, I needed apples and sugar, raisins and spice.
Today, I needed a pie.

Ouita Michel
Around the Table

THURSDAY, JULY 28, 2022, KNIFE IN HAND, I WAS SWEATING IT out in the prep kitchen at Fasig-Tipton, getting ready for CRAVE, the big music and food festival in Lexington. I hadn't heard much about the flooding in eastern Kentucky, but later that day, the news started to spread.

I texted my friend Kristin Smith, a chef and restaurateur in Corbin, Kentucky, to make sure she and her wife, Mel, were OK. The last flood had hit their farm hard, and I started to worry. Kristin texted back, "We are good. The flooding hasn't been as bad at the farm. We were in Hindman for the Appalachian Writers' Workshop tho last night when it hit, and it's one of the most traumatic experiences of my life. We were able to get out by this afternoon. It's bad over there. Real bad."

My stomach flipped. I couldn't leave Lexington until Monday. The next day, Kristin and Mel headed back to the school to see how they could help. Katie Startzman, a fellow chef and restaurateur from Berea, joined them. Over the next two weeks, thirty volunteer chefs from central Kentucky would travel to Hindman Settlement School to help cook for folks in need. Even more chefs volunteered with World Central Kitchen, and more still at the CANE Community Kitchen in Whitesburg, Kentucky.

Hindman Settlement School is one of my favorite places to cook. The kitchen is large and very well equipped, with a great porch across the front filled with a picnic table and several rocking chairs for visiting. For several years, I had coordinated a "Dumplin's and Dancin'" dinner there—celebrating the culinary heritage of eastern Kentucky and the connections between our Kentucky foodways and our musical and literary traditions—along with chefs, farmers, and others from around the region.

It's not every place you can cook up a pot of leather britches and pans full of corn bread, dumplings, cabbage, and pork roast and then square dance the night away after the dishes have been put up and the dining room cleaned. Diane Owens and Vivian Richie run the kitchen. They are wonderful to work with and make us feel welcome, with a kind word for everyone. We work really well together, and they are always ready to jump on any task to help. Over the years, we have become cooking buddies, and one year pre-pandemic, they came to see my kitchens in Lexington.

I was worried for them and for the school.

Troublesome Creek had roared through the first floor of the school's main building and several smaller classroom buildings, destroying its new offices, archives, laundry facilities, and guest rooms. Thankfully, the water did not get to the higher floor, and the kitchen, dining room, and large gathering space where we often square danced after dinner were intact.

Those first days after the flood, though, there was no power and no running water. Kristin and Katie had set up big flat grills, along with gas grills, out on the front porch. Diane, Kristin, and Katie cooked there and used gallon jugs of potable water to clean cooking utensils and pots. Folks displaced from their homes in the community were relocated to the school's residential rooms up on the hill. At one point, forty-five people had taken up residence at the school, including several children. They had lost everything.

That weekend, I was knee-deep in festival cooking and madly trying to get water and other supplies to Diane, Vivian, Kristin, and Katie. Fellow board members bore the brunt of soliciting donations and filling truckloads of water, sheets, cleaning supplies, and much more for transport to the school.

Chef Amy Brandwein from Washington, DC, connected us to World Central Kitchen. They were on the ground that first weekend, setting up a massive meal distribution. Hindman Settlement School could be one of their distribution points, but it would take a couple of days to get everything established. They would deliver twice a day, enough to feed one hundred people a hot lunch and later a hot dinner. Diane and Vivian would focus on breakfast.

Ouita Michel

On Monday, August 1, after what seemed like an eternity, I finally headed to Hindman. I took my daughter Willa and fellow chef Daniela Gonzales Garcia, visiting from the Lee Initiative. Power had been restored to the school, but water still had to be boiled before use or brought in by the jugful. The damage to the school and the destruction Troublesome Creek had wrought on the community of Hindman were shocking. It took a few hours for the magnitude of the flood and the losses the community faced to dawn on us. Even today, the losses continue to unfold, and I don't think anyone has been able to fully comprehend the scale of community challenge and change in eastern Kentucky.

When we arrived that Monday morning, Diane was finishing up some breakfast on the outdoor griddles and looked exhausted. Cooking a hot breakfast for everyone was one of her top priorities. I sent her home to rest, and we started cleaning and organizing. In the drive, a large portable water tank called a buffalo was connected to a garden hose, which ran through the nearest kitchen window into a small sink. We used the stove to boil water for dishwashing.

The Great Hall in the Mike Mullins Center had been set up as a distribution point for household articles donated for those who had lost everything. Personal hygiene products, blankets, sheets, pots and pans, clothes, even generators—it was all there. A mountain of disposable plates, to-go boxes, napkins, and cups accumulated in the dining room, along with all kinds of packaged foods for folks to take back for snacks to keep their energy up while they worked to dig out.

We decided, given the water situation, that we would keep cooking to a minimum and focus instead on keeping things as clean and sanitary as possible. That meant mopping the dining room and kitchen floors and sanitizing all surfaces multiple times per day, cleaning the bathrooms, organizing the kitchen, wiping all surfaces possible. Mopping floors felt like an unending Zen exercise. The fine dust and silt that settled on everything streaked and tracked when mopped, but eventually the dirt gave in, and we washed it away.

World Central Kitchen brought our first lunch that day, and we had a big sign broadcasting "Hot Meals." People came and sat in the coolness of the air-conditioning to gather their thoughts. Many took their meals to go, preferring to head right back to their homes, where many family members were also waiting for a meal. Transportation was a luxury because at first gas was hard to come by and many roads were impassable.

Small things live in my memory the most. We brought ice-cold watermelon with us that first Monday, and people devoured it, especially the kids. The flood had disrupted food distribution in the area, and relief cooking focused on bulk hot food.

Raw fruits and vegetables were in short supply. Many folks who relied on gardens for fresh foods had seen their crops destroyed or contaminated by the flood. The lucky ones whose gardens survived went to work harvesting and donating what they could—tomatoes, summer squash, peppers, melons, beans, and even fresh basil. We cut up all the fresh fruit we could, and I think almost every meal featured pickled onions and cucumbers, as well as sliced tomatoes.

Kelsey Cloonan, a member of the settlement school staff, even arranged for additional donations of produce to establish a free community farmers' market and to compensate the school's farmers' market vendors for their losses.

The guests who came to eat in the dining room were so polite that they wouldn't approach the buffet unless invited, and then they carefully asked if they could take more than one meal. I had to remind myself to invite everyone to come forward—and to help themselves to as much as they needed for the ones at home as well. One lady wore her leopard-print pajamas into the dining room. They were what she had on when the flood came; she had lost everything else. All available housing was given over to those in need; we volunteers drove in and out each day so we wouldn't be an additional burden to the community. That long road through Jackson, with rubble piled so high on both sides, was very dark and quiet on that first night heading back to Midway.

In the face of a destructive event as large as this flood, it can feel daunting to volunteer. What difference could we make? Still, we wrote a schedule for volunteer chefs to come at lunch, dinner, or both—to serve the meal, clean, and make folks feel welcome. We wrote a simple set of rules for kitchen operations, and as hard as communications were in the face of the sketchy internet and a big circle of cell phones, people showed up. They brought food, they washed dishes and cleaned tables, and they helped. Katie Startzman, especially, helped through to the very end of meal production.

As those first few weeks passed, gas and water were restored. The need for multiple daily hot meals declined as families were relocated to FEMA trailers and began to leave the school. World Central Kitchens packed up their operations. The Hindman Settlement School kept providing lunch but scaled back its dining room operations. Folks could still come and get things they needed from the large supply of household goods stored at the school.

After a few months, the school focused its cooking efforts on bringing members of the community together for healing and rebuilding with a new program called Gather & Grow. As a member of the Kentucky chef community, I learned that we really do show up for one another and that next time we'll be even more adept and ready to dive in than we were this time.

Many droplets make an ocean, and it felt like the collective effort of volunteering made a difference in the aftermath of the flood. Even better, we connected to the community of Hindman, and folks who had never been to or heard of the school connected to its mission and to the folks who work there. I am so proud of the efforts of every member of the Hindman Settlement School staff and the hard work they have done to rebuild the school and its amazing programs. It's still one of my very favorite places to cook, and I hope that the next time I have the privilege of making a meal there, square dancing will follow.

Jane Hicks

Mr. Still's Hat

Mr. Still's hat
might well be the straw crown of a gone
and benevolent prince—retrieved
in sorting ruined archives
after a mighty tide of destruction.

He wore it when I last talked
to him that August day, rare occasion
when I found him alone, porch-sitting.
"Look there, Tennessee."
(He called me Tennessee,
whether he didn't know my name
after twenty years at the workshop
or he let me know he remembered that fact.)
"He's quiled up."
The *he* being a fat copperhead
abask in the sun. Mr. Still pointed a cane
across the parking lot shimmer
to a sandstone, very near the serpent color.
From that cue, we talked snake lore
across cultures, especially the Maya,
a topic of mutual interest. He told me
of a trip where he saw the pyramids of
the serpent god, Kukulkan,

slithering in solstice shadows.
We digressed to a story of
house cleaners finding his childhood
shirt in the bottom of a trunk.
A fleur-de-lis pattern
he thought to be flowers and buried
it deep. I told him the fleur-de-lis was once
called Napoleon's bees. Maybe
a boy would like such. Then we spun
bee lore: souls, messengers,
familiars, and signs. We quoted Whittier
as the dinner bell rang and we rose to go.

I wrote that poem; wondered if anyone
told the bees when they buried him
the next spring. They laid him
high above the creek fork—far above
the most troublesome tide or fierce flood—
on a May Day, when children once
danced about a Maypole on those grounds.

On the third day of mud, muck, and sorrow,
they found his hat, intact. A blessing
from one who watched, waited, wary
of troubles and tides, a believer
in messages and signs.

Erin Miller Reid
What We Saved

I WONDER HOW MANY TIMES TROUBLESOME CREEK FLOODED IN those first years after Katherine Pettit and May Stone founded Hindman Settlement School over a century ago. How many times, as they cleaned up debris borne on the back of turbulent waters, did they shake their heads at each other in dismay and acknowledge that the creek was aptly named? Did they ever wish they'd set up their school somewhere else and taken their cause to higher ground?

Despite the ever-present threat of floodwater, I am thankful Pettit and Stone stayed on the banks of the Troublesome. Hindman Settlement School is one of the dearest places in the world to me. Silas House has called it a *thin place*, where the ethereal transfuses the physical. Maybe it's the lick of the creek over mud-slick stones, the way the campus is hemmed in by steep ridges green with the climb of kudzu, the vapored rise of morning fog from the mountaintops. Maybe the air is suffused with the founding mothers' fortitude and optimism. Maybe the words of resident writers James Still, Lucy Furman, and Albert Stewart still fertilize the soil. Whatever the reason, I agree: Hindman Settlement School is a place where the distance between heaven and earth disappears.

I first arrived on campus in 2018 for the Appalachian Writers' Workshop. I've attended the weeklong event each year since, including the virtual workshop during the height of the pandemic in 2020. Hindman is the place that immersed me in the heart of Appalachian literature.

Erin Miller Reid

It's where I've honed my own writing craft, and I had my very first mandolin lesson on the porch of the Mullins Center. It's the place where I've forged friendships that feel more like family over plates of tomato pie and bags of Grippo's chips, over tarot readings and late-night visits on the front porch of the Stucky house. It's the first place, at thirty-eight years old, that I ever felt proud of my eastern Kentucky accent. It's where I slept on July 27, 2022, the night it rained without relent and the waters of Troublesome Creek overflowed to historic levels.

When the sun rose the morning after, bleary-eyed members of the Appalachian Writers' Workshop, many still clad in pajamas, surveyed the damage. Little about the place resembled heaven. Landscaped bushes had been uprooted and dropped on their sides. The smell of gasoline hung in the air. A soiled couch cushion lodged twelve feet high in a tree on the creek bank. Water submerged the iconic Hindman Bridge. Matted leaves clogged its metal suspensions. Both the historic Uncle Sol's Cabin, filled with musical instruments to teach local children, and the James Still Classroom Building, home of the settlement school's dyslexia program, swam underwater. The first floor of the school's main building, the Mullins Center, was flooded as well, which meant the destruction of the administrative offices, two apartments, and, worst of all, the archive room, which stored irreplaceable books, letters, photographs, and artifacts from the school's history.

Without power or water and with more rain in the forecast, workshop members evacuated campus as soon as roads were passable. As I navigated muddy rockslides, downed trees, and roads so flooded I was forced to drive on the rumble strip, the drowned archives haunted me. If only we'd known in those hours before the flood, as we relaxed on a patio just yards away from the archive room, that so much would be lost by morning, we could have rescued those priceless objects. Dozens of us could have carried folders and boxes to higher ground. History could have been saved.

If only we'd known.

Decades before, former director of Hindman Settlement School Mike Mullins had known. He'd recognized the value of preserving the memorabilia that formed the school's history. Early in his tenure, he procured grants to aid in the preservation of the archives through microfilm.

Thanks to his foresight, copied records through the 1980s reside on the campus of Berea College, but on that July 2022 night, the originals were decimated.

Two days after the flood, I heeded an online plea for help to salvage the archives. After urgent phone calls to anyone I thought might be able to assist, I reached Jeremy Smith, director of East Tennessee State University's (ETSU) Archives of Appalachia. He generously agreed to make space in the university's food-service freezers for waterlogged documents, an effort that would halt the growth of mold until further assessment and restoration could begin.

I recruited a friend with a truck, Scott Honeycutt, language and literature faculty from ETSU, to haul the documents, and we drove two hours from northeast Tennessee to Hindman. The sun was blindingly bright and the sky clear blue, in sharp contrast to the dreary morning of the flood, though the roads were still smeared with mud and littered with broken limbs. We reached campus by lunchtime and found it transformed. Once the water receded, Hindman Settlement School instinctually pivoted and became a hub of outreach for flood victims. Bottled water towered on the main building's front porch, where we'd snapped beans just days before. The Great Hall, once filled with poetry, prose, and contra dancing, now brimmed with paper towels, gallons of bleach, and heavy-duty trash bags. I looked up the hill toward the Stucky house. Two nights ago, I'd sat there with friends in worried silence. Now a family of houseless flood refugees inhabited that quiet unease.

The lower-level archive room was pitch dark except for the light cast by a pair of camping lanterns. My rubber boots, the ones decorated with cheerful bumblebees that I had bought to garden in, the ones that would later deteriorate from the unknown chemicals that imbued the floodwaters, squelched in sour, inch-thick mud. Gnats swarmed above file cabinets. Labeled drawers were filled with contents so waterlogged they swelled and wedged in place. When we managed to dislodge the items, we found advertisement pamphlets from 1975, a playscript from 1959, old ballads recorded in blocked typewriter font, and tunes written out in shape notes. We found black-and-white photographs of teenagers square-dancing, school children reading on their bellies beside the Knott County Bookmobile, and adolescent boys chopping wood and fishing in the creek with trousers rolled to the knees. We found daguerreotype portraits in a Victorian album, James Still's bent hat, and a leather-bound ledger printed with May Stone's meticulous hand.

The task of saving it all, drying it, and cleaning off the mire seemed impossible. Even with hard-working volunteers unloading drawer after drawer that Saturday afternoon, progress was slow and unfulfilling. Jeremy Smith from ETSU advised me, "Focus on what you think is unique to the settlement school, what can't be found anywhere else."

Books on the bottom shelves at one end of the room looked irreparable. Their spines were twisted and pages clumped with khaki-colored mud. The covers were unreadable. Were they worth saving? My first impulse was no, forget them. Author Robert Gipe, who also helped in the archive room that Saturday afternoon, said, "Someone thought these books were important enough to keep."

I thought of my own kids and what I had saved from their childhoods. The baby teeth kept in a mercury glass canister; crinkled smears of finger paint on construction paper; kindergarten worksheets printed with giant, lopsided signatures; report cards; hundreds of baby photos that attempt to preserve each spit-bubble smile, each oval-mouthed coo; the outfits they wore home from the hospital; a pair of teensy newborn socks so I won't forget how tiny they once were.

I keep it all in a rubber tote. By any other person's standards, the trinkets might appear meaningless. Some might even call them junk. But to me, these are the treasures of motherhood, the ephemera of my children's lives. These items are invaluable to me.

I reevaluated the muddy books. Someone long ago thought these books were important, worthy enough to keep. But, unlike my tote of childhood memorabilia, the archives at Hindman Settlement School are so much more than one family's legacy. The archives chronicle the past of an entire region. Robert was right about the books' worth. It turns out they belonged to the library of Elizabeth Watts, an early teacher and the school's second director. She's buried on a campus hill next to James Still. For all I knew, the title pages were bedecked with the autographed inscriptions of Harriette Arnow, Wilma Dykeman, and Verna Mae Slone, authors foundational to Appalachian literature. Those books were precious too.

Help did not only come from ETSU's Archives of Appalachia. In the effort to save the archives, Melissa Helton, the current literary arts director at Hindman Settlement School, coordinated with multiple other volunteers and organizations, including Kentucky Natural Lands Trust and archivists Alexia Ault from Southeast Kentucky Community and Technical College and Lori Meyers-Steele from Berea College. Clem's Refrigerated Foods stored rescued materials for months in Lexington, Kentucky, at no cost to the school, and Lydia Kitts, a fabric conservator, restored Mr. Still's hat. Volunteers from the Appalachian Writers' Workshop, the Daughters of the American Revolution, and the local community dedicated hours to sorting and drying photos and documents. Days after the flood, the individual, hundred-year-old pages of May Stone and Katherine Pettit's handwritten account of the school's origin lay in rows on the wooden floor of the Great Hall. Photographs and 35mm Kodak slides were clothespinned to twine stretched between cafeteria chairs in a section of the Great Hall that had been cordoned off from the rows of folding tables piled high with cleaning supplies. In black marker on an empty manila folder, a volunteer had made a sign for the corner: "Working hard to save our heritage." Once the archives' rescue corner was dismantled weeks later, Melissa saved this folder to be officially added into the collection as the first new item postflood.

Already, my recollection of the flood has faded. A single brown dress shoe washed onto the highway. Silt-filled dulcimer bellies. Pine trees bent like overused bottle brushes. Concrete driveways reduced to rubble. The roof of a trailer lit up in orange flames. The lifeless, swollen body of a small dog between the center yellow lines. Folks piggybacked on ATVs out to survey the aftermath. Porch steps leading to nowhere. Bright blue baby pools high in tree branches. An upturned set of empty kitchen cabinets. A white plastic lawn chair resting on the side of the road.

The images bob in and out of my memory like flotsam on a river, visible one minute and submerged the next, dislodged and transient. The reason I filled my kids' baby books was so I'd remember. I didn't want to forget first words, first birthday parties, milestones like smiling and rolling over. I wrote it all down. The founders and directors of Hindman Settlement School did the same. They saved each school brochure, each tape-recorded oral history, each letter, each graduation program, and each church recipe book to remember. It's the act of preserving what matters to us that anchors us.

I remember my mother telling me my namesakes were my grandma and an aunt. I remember when my dad told me we were mostly from German stock. On my mom's side, we descended from Bulgarians and Italians. Whenever he had the chance, my grandpa told tales of being raised with twelve siblings, saying they would have been fifteen if it hadn't been for the flu. My grandmother scribbled down her mother's recipe for pasta fagioli, and my adopted grandmother taught me how to fix corn bread in a skillet.

My personal history is important to me and my family. I tell it to my children in hopes that they will relay it to their own kids someday. I have the same hope for our Appalachian heritage, that it will be passed to future generations. Naturally, we want to share our stories and traditions, but if we aren't the ones to remember and to sustain our own legacy, who will?

In a remarkable yellow-tinged, pocket-size photograph that was saved from the archives, a woman wearing a black bonnet and a long, checked skirt, perhaps Stone or Pettit, crosses the flooded Troublesome Creek using a row of wooden chairs. She stands on two center chairs frozen in motion as she works to reposition the rear chair to the front so she can step forward. She's building a bridge across the high waters, a fitting metaphor for the archives: taking what's behind us and moving it forward.

Someday it will be our written accounts, filmed interviews, and social media posts telling future generations how whole communities rolled up their sleeves and shoveled mud from living rooms, how we fried bacon and stirred gravy on propane grills, how truckloads of supplies poured in from hundreds of miles away, how we laid out soggy, ink-smudged pages to dry, how we cried on each other's shoulders for the greatest loss, human life. Then, those future people will learn how we all pulled together, saved the past, and pressed forward, just like the woman crossing the swollen creek with ladder-back chairs.

We will tell our story. Mothers to sons. Fathers to daughters. Parents to offspring. The archived pages and recordings to anyone who will read and listen. For we are the children of these mountains, the folks of these hills, and the stewards of our history.

Elizabeth Lane Glass
Aching for Troublesome

THE MORNING AFTER THE FLOOD, I WOKE IN LOUISVILLE WISHING I were in Hindman. Dreading that I had to face teaching English classes instead of writing and being with people from the Appalachian Writers' Workshop (AWW), the folks who are *my people*, I logged onto Facebook. Instead of seeing the pictures of found treasures, folks writing on laptops and in notebooks, people gathered in the evening, I saw the flood. I pored over the news, the posts, everything I could find.

I cried the whole way to the University of Louisville, then spent the first twenty minutes of each class telling my students about the flood.

"You know the writing workshop I love to attend? The one going on—that *was* going on—this week? Look at these." I pulled up Facebook and showed them videos and photos from Melissa Helton's growing collection.

"Remember the essays we've read about mountaintop removal?" I asked. "Recall the flooding we've talked about?" I'd shown my English composition students videos of mountains being blown up, of folks from the coalfields talking about mountaintop removal, of people discussing the flooding that affected their lands every year because the creeks have nowhere to go once the strip-mining debris, the overfill, covered the creek beds. We'd read articles about these issues when I taught them how to write op-eds. The semester was ending soon, so they knew these things. They were moved that morning after the flood, more than many of my students had been, for whom the situation was less tangible, less immediate.

"I want to help," I said, trailing off.

I got home and walked my dog, my mind solely on the flood. I ached with not having been there—to what? To "help"? But what help would I have been, with my walker and cane? How much help would I have needed if I hadn't stayed in Preece, way up on the hill? I'd planned to ask to stay in that building if I'd been able to apply instead of teaching. I'd have *needed* far more help, and *been* of far less help, because: disabled.

I spent more time on Facebook in the weeks after the flood than I have in years. I hurt for the settlement school and for my friends and future friends. I ached for the lost cars and books and belongings. For the buildings I love so much. For . . . everything.

I wanted to go to Hindman, pass out goods to those affected by the flood, but being from Louisville, I'd have needed a place to stay. Folks who *really* needed the space to stay in because they lost their homes were staying on campus and in the nearby hotels.

When the archive work at the settlement school was going on, I wanted to go. I thought, "I could do that," but again, I would need somewhere to stay. I realized my dog wouldn't be able to go too, which came to money because I couldn't afford to board him or pay someone to come in and care for him. But mostly I was afraid I would be in the way. To get around campus, I would need my walker. The kind I had meant I couldn't carry anything on my own, so someone would have had to help me. I wanted so badly to help, but I would have been more of a hindrance than a help. Because: disabled.

I did what I could. I donated money. My birthday is in August, so I did a Facebook birthday fundraiser for the settlement school and raised far more than I could have donated myself. I told everyone I could about strip mining, about the flood. I've taught about Appalachia in all of my classes since 2016, if not before. I'm teaching Appalachian studies this coming fall semester and have also done so at Bellarmine University in Louisville. In addition to the three novels and collection of essays by Appalachian writers we're reading, I'll spend time talking about the mountains, strip mining, job loss, flooding. I'll show videos and play music by Appalachians, making things visual and audial so the students will appreciate Appalachia on as deep a level as possible while not being there.

People with disabilities face all sorts of issues when it comes to natural disasters like the flood in July. Despite the No One Left Behind policy developed post-Katrina to include folks with disabilities and those who are older in emergency planning, in actuality, this still doesn't happen. It was developed because Katrina killed at least 1,330 people, of which disabled and older folks were frequently hardest hit. In Louisiana, 71 percent of those who died during Katrina were over sixty, and people with disabilities—especially with the intersection of poverty—were often left behind during the evacuation.[1]

An article in the *New York Times* on July 31, 2022, profiled the people who died in the flooding in Hindman and nearby areas. In this article, everyone, except one family, was elderly, many of whom also had co-occurring disabilities. At least thirty-seven died in the flood in eastern Kentucky, and the elderly and disabled were especially affected—again, especially if poverty was intersected with age or disability.[2]

After evacuation during a flood or other natural disaster, we disabled folks are also more likely to face disruption to services—medical, carers, and others—necessary to daily living. Often, there are few or no emergency personnel able to evacuate us, particularly if we use power wheelchairs—or any mobility device. We lose these mobility devices, and there are frequently no accessible shelters available during an evacuation. When we *can* evacuate, we often lose our medications in the disaster, and because we lose access to services, we frequently can't get them replaced in a timely manner.[3]

I imagine what would have happened during AWW if I were not staying in Preece at the top of the hill. I might have lost my medications, including the ones that help me walk. It would have been hard to fit my walker in someone's car to go up the hill, and it would have been impossible if other people were being transported at the same time. I would have been a burden in the flood had I been at Hindman. It's humbling to realize.

I had an evacuation plan at home when my roommate and I were partners. Now that I'm on my own, I naively—and foolishly—don't have one. I'm not sure if I'd even have the foresight to grab my medications if evacuating for a natural disaster. It was an issue for elderly and disabled folks after the flood in July too.[4]

I've taught another semester about mountaintop removal, about the flood. Like every other semester, students—nearly all of them—said, "*How* did I not know about this?"

How indeed.

I tell them how. I tell them exactly how: because, like people ignore and stigmatize those with disabilities, people ignore and stigmatize Appalachia and Appalachians. This is nothing new to anyone reading this, but it makes my students think. They had done this. Or they were from Appalachia and were glad I talked about it because they felt seen.

When I taught Cultures of America, I included a section on Southern Appalachia. Obviously in any culture, there are a million cultures, but many truths can be learned from reading primary texts by Appalachians, like those by my friends from AWW—Silas House, Jason Howard, David Joy, Leah Hampton, Robert Gipe, Carter Sickels, Crystal Wilkinson, and Annette Saunooke Clapsaddle—and by Elizabeth Catte, whom I don't know but who has an important voice. These folks write in ways that capture a part of where they're from and in ways that are both universal and specific. So many, many friends from AWW do the same. So many, many friends from AWW are incredible writers, and I wish I could teach them all.

I hope that this book can be used in classrooms so more students learn of the flood. About mountaintop removal and the problems with overfill covering creeks. About the pride and love those of us who return to the settlement school year after year have for the area.

And so it's time to apply to the AWW again. I don't know my summer plans job-wise, but I know I need to attend if I can. I know that despite my walker, now a rollator that is big for the dining area, for the classrooms—for everywhere—I will go. Even though: disabled, I'll apply. I ache every day the workshop is occurring if it's a year I'm not there. I imagine the new friends I'll meet, the amount of material I'll learn, the things I'll read and hear, the keynote I'll absorb. The knowledge I'll gain that I can then share with my students.

Even though I'm not from Hindman, in my heart, it's home. Every year, returning to AWW at the settlement school is going home. And I need to go home. The settlement school is where this disabled girl's heart lies. In the chapel, in the woods, in the buildings, but especially on the banks of Troublesome, which I can get to again thanks to my big ole rollator.

Because people with disabilities are part of this family too.

TROUBLESOME RISING

Road north out of Hindman toward KY 80. © Julie Rae Powers

Notes

1. Kathleen Otte, "No One Left Behind: Including Older Adults and People with Disabilities in Emergency Planning," ACL Administration for Community Living, September 4, 2015, https://acl.gov/news-and-events/acl-blog/no-one-left-behind-including-older-adults-and-people-disabilities.

2. Shawn Hubler et al., "What We Know about the Victims in the Kentucky Flooding," *New York Times*, August 1, 2022, https://www.nytimes.com/2022/07/31/us/kentucky-floods-victims.html.

3. Jodie Bailie et al., "Exposure to Risk and Experiences of River Flooding for People with Disability and Carers in Rural Australia: A Cross-sectional Survey," *BMJ Open*, August 1, 2022, https://bmjopen.bmj.com/content/12/8/e056210; Laura M. Stough and Ilan Kelman, "People with Disabilities and Disasters," in *Handbook of Disaster Research*, ed. Rodríguez Havidán, William Donner, and Joseph E. Trainor, 225–42 (Cham: Springer, 2018).

4. Shawn Hubler, "What the Floods Left Behind: A Devastated Kentucky Views the Damage," *New York Times*, August 1, 2022, https://www.nytimes.com/2022/07/31/us/kentucky-flood-photos.html.

Kari Gunter-Seymour
Bluegrass Navy

Destruction wasn't the only story,
though eastern Kentucky
did have a catastrophic collapse—
water as high as 14.5 feet above flood stage.
There were graphic images plastered
across news outlets and national figures
flash danced for the cameras.

Meanwhile, Kentuckians rescued
their own selves—truckers, teachers, nurses,
preachers—generations of bluegrass born
arrived from all parts of the state and nation,
wearing all-weather hunting gear,
towing bass boats, driving Fords and Chevys
filled with towropes, chain saws,
beef jerky, bottled water, pet food.

Folks spontaneously set up grills
in Walmart parking lots to fill empty bellies,
kinship strong as the granite and iron
from which the Appalachians
themselves were formed,

TROUBLESOME RISING

showed the rest of the world how it's done—
in johnboats, on Jet Skis, inside rescue centers
and emergency shelters,
no sleep for days, looking after their own.

Courtney Lucas
Mud: A Story

AFTER THE FLOOD, HE LEFT KNOTT COUNTY, BUT HE DIDN'T GO far. He thought Pike County would be a good enough distance—an hour or so away but still home, still too entrenched in the muddy aftermath to be concerned about a missing man from two counties over. They had their own missing to find.

He quietly slipped into donation drop-off points, took some food and a change or two of clothing, and left without saying a word more than a muttered "Thanks." To fill the days, he drove around the area—Whitesburg, Fleming-Neon, Virgie, Dorton—and grabbed a shovel here and there, helping people he didn't know clear away the mud and debris from whatever remained of their homes. He moved through communities where everyone had the same wide eyes and lips pressed into thin lines, communities where the residents tenderly scraped mud from wedding pictures in broken frames. They weren't in a position to question the help they received.

One evening, on his way back to the county elementary school now serving as a donation center and shelter, he passed a hastily constructed sign scrawled in black marker on a piece of soggy cardboard: "WE ARE RESILIENT. WE WILL RISE." The word *resilient* stuck in his craw like something dry. Last time he checked, no one had asked him if he wanted to be resilient. Like he was some kind of spider in the bathtub that climbed back up every time he got washed down.

It was nearly dark by the time he'd eaten his dinner, and he wanted nothing more than to be alone, to just sleep without someone looking at him like he was some lost puppy. He'd noticed the looks he got, the poorly hidden stares of those who wondered about the young man who never said a word and didn't appear to know anyone in the community, despite returning every evening with muddied boots and clothes. It was only a matter of time before someone asked where he came from outright, and what would he say then?

He climbed into his truck and reclined the seat as far as it would go. The streetlight above his truck flickered and sizzled before going out, and he was back in his buddy's place, back with the taste of cinnamon-apple moonshine in his mouth, too drunk to feel the burning in his throat.

No one says anything at first, and with the power out and the radio silent now, the only sound is the rain pounding the roof, the windows, and any poor soul unlucky enough to be caught outside.

Finally, one of the men speaks: "Rain's got the lines down, I bet."

Then, the silence now broken, another speaks: "Tree fell or something."

A third: "Guess it's time to head home."

But no one moves. Everyone sits in the darkness, bleary eyed and grinning, letting the warmth settle in their stomachs.

And then he remembers his grandmother, sleeping at home, jolted awake and suffocating, her oxygen concentrator sitting idle without power, unable to help her breathe, and his grandfather, down in his back and all of his bones, lacking the strength to pull out the generator.

"Mamaw," he says, lurching forward too quickly, sending the world spinning. The darkness makes his dizziness worse, disorienting him in the once-familiar space. He shuffles in what he thinks is the direction of the door, arms outstretched; stubs his toe on the table leg; curses; but finds the cool of the doorknob, twists, and pulls.

The rain is unrelenting. It takes only moments for him to be soaked through to the bone, clothing sagging from shoulders and hips, weighing down his already unstable movements. He thinks he hears one of his buddies shout at him, but he can't make out the words for the rain and the roar of the swollen creek.

He sloshes through a few inches of water, and it drenches his tennis shoes, soaking his socks and feet, nearly causing him to lose his balance. Even in the heat of summer, the water chills. Then the water is over his shoes, rushing around his ankles, his midcalves. The lightning flashes, illuminating the darkness of a night without electricity, and he realizes that he is not walking through runoff; he is walking through the overflown creek, rising to a place where it shouldn't be.

The water moves so fast, and a bitter taste rises in the back of his throat. The creek shouldn't be here. It's not supposed to be here. It's the only thing he can think as he turns and walks in the other direction. It will take longer to get home this way, but at least he'll be on higher ground, and he should be able to cross to the other side of the creek. But what should and shouldn't be is all mixed up in his mind, all muddy and swirled around, and he sinks to his knees and throws up into the water.

After his mother had come home, seventeen years old and pregnant, after she had refused to tell the name of the boy who'd done it to her, after she finally left for good the February before his seventh birthday, his grandmother took her place, becoming the only real mother he'd ever known. And now she is choking on the fluid in her lungs, unable to pull a breath deep enough to get it up once and for all, making do with spurts of short, hacking coughs.

He coughs once and wipes his mouth with the back of his wet hand. "Mamaw," he says as he stands to his knees and trudges through the rising water with new resolve. The lightning is frequent enough to more or less light the way, but there are a few times when his foot slides or he kicks something hard under the water. His grandmother had always wanted him to put down the bottle, to go to church with her on Sundays, but he never was the faithful type; he likes to believe in what he can see. But now, as the waters rise all around him, he tries to pray. He tries to pray to the Lord to let him get home, to let his grandparents be okay, and to just let him get to them, but all he can say is, "Mamaw."

When he reaches the bridge, he feels that God has heard his unspoken prayers anyway, because the water is still a few feet below the bottom of the bridge. He peers over the railing, and the lighting flashes, revealing a muddy brown whirlpool carrying a silver sedan like a Tinkertoy. But it's dark when the water smashes the vehicle into one of the bridge supports, shaking the bridge and almost throwing him from his feet.

He quickens his pace, nearly running, feet now on higher ground but still squelching water from his soggy socks and shoes. He knows the route by heart, enough to make it to the mouth of his grandparents' holler, tripping only a handful of times even in the near-total darkness.

There, just ahead, a row of pale blue lights glows a few inches off the ground—his grandmother's solar lights staked into the grass along the front walk of their house, like a beacon of dry land; the water hasn't gotten high enough yet to wash them away. He runs toward those lights like he's running toward salvation—not his but his grandparents'—until he again runs through creek where creek shouldn't be. A wide stream of water gushes out of the holler, coming dangerously close to the house but even closer to his truck.

His grandparents just bought him that truck last month even though he knows they couldn't really afford it. He'd never driven a brand-new vehicle before—this is his first—and they got it to thank him for all he does for them, for helping out around the house and picking up their groceries and making sure they take their medicine, all things he does without thinking, things he never considered not doing.

He throws open the door of the truck and hops inside, turns the keys already in the ignition, but isn't sure where to go. He has to get the truck to higher ground, but he can't think of a place untouched by water, and he wishes he was sober, or at least less drunk. He drums his fingers on the steering wheel to calm his nerves and remembers Sheriff Slone's house on a hill nearby. If the water gets that high, nothing could save them but an ark, and surely the sheriff has other things to worry about than a drunk driver getting behind the wheel only to move his truck to higher ground.

The headlights cut through the rain still coming down in sheets, and he parks in the sheriff's driveway, right next to the darkened house, figuring they will understand once the sun rises. The truck should be safe up here on the hill.

And from the hill, he sees.

Lightning flash: His grandparents' house. The house he'd been raised in. The only home he'd ever known.

Lightning flash: An empty foundation.

Lightning flash: The house, already yards away, carried by the river, spilling clothing and tchotchkes and memories out the broken windows.

Lightning flash: A splintered roof caught in the branches of a half-submerged sycamore, dangling shingles and the butterfly quilt from his grandmother's bed.

Lightning flash: Nothing.

He woke the next morning just as the sky began to lighten. Condensation fogged the windows of the truck, and drops of water, overcome by gravity, slid down the wet glass, clearing a little trail. For a moment after wiping the window, he was confused to see the elementary school.

They were already serving breakfast in the cafeteria—big trays of watery scrambled eggs, too-crisp bacon, and dry biscuits that had been left in the oven just a little too long. He ate in silence.

A man and a woman conversed at a nearby table.

"You know the Taylors that live up around Little Creek? Real big house?" the woman said.

"I don't reckon."

"They got them zebras over there."

"Yeah, yeah, I know the ones."

"All but one of them things got washed away. Somebody found one dead up in the Ohio River, someplace near Ashland."

"I wouldn't've thought stuff could've washed that far."

"Me neither."

They were silent for a moment.

"All them missing people," the man said. "Who knows where they'll start turning up?"

He'd known it was only a matter of time before search areas were widened, before he'd have to move on from Pike County, just like he'd moved on from Knott County. He couldn't explain why he hadn't been home, why he'd chosen to move the truck first, why he wished more than anything that he'd been in that house too. He thought of the zebra, its striped corpse bloated with fetid water, the shock on the face of the person who found it. He wondered where they'd find his grandparents, where they'd be, and what kind of shape their bodies would be in. Maybe they'd found them already, or maybe they never would.

He finished his breakfast and got back into his truck, wondering just how far he'd have to drive to escape the mud and the water, wondering if any distance would be far enough.

Silas House
Pulled from the Flood

For long stretches we drive through people's belongings on either side of us. We have to creep along because the debris is close. To navigate our way through, we are required to see. I glimpse a blue-and-red plastic child's bed that is shaped like a race car, a water heater, a box of Little Debbie Nutty Buddy bars, a baby's nightgown, one shiny red high heel. A couch stands on its end. A large portrait shows a woman with two children smiling in matching white sweaters behind a shattered pane of glass. In the trees hang strips of unidentifiable plastic, tufts of pink fiberglass, jagged yellow sheets of plywood.

The flood hit eastern Kentucky three days before, on July 28, and this section of Highway 15 between Jackson and Hazard only recently became passable. The creek has receded, but it still runs with a muddy, ferocious determination, slithering around a car lodged upside down between two boulders, a large toolshed that has somehow remained intact, a French door, a mess of broken furniture. Washed onto the higher bank, a mobile home bends like a loaf of bread that has been mashed in the middle. Then, a brick house with half its bricks fallen away. Another small house has been washed ten yards away from its foundation. All of the windows have been busted out, either by trees, detritus, or the weight of the rushing water.

A line of soldiers outfitted in grayish camo march down the road in single file. A woman in an orange vest directs traffic around a pile of broken lumber and twisted metal that hasn't been moved out of the road yet. On a side road, one bridge is completely washed away. On another bridge, an entire house lies on its side, its floor beams exposed.

Trees are uprooted and leaning all along the sides of the road, their bark stripped away to reveal startling whites and yellows. Telephone poles point in forty-five-degree angles. Waterfalls rush out of the mountainsides, and in some stretches our road has partly collapsed, so it is reduced to one lane.

Now we come to a wide, level section that I realize has recently been made that way; the emboldened creek has acted like a bulldozer cutting a swath through here, leaving in its wake more personal items. People move cautiously through the debris, bent over as if looking for Easter eggs. But instead they're searching for moments from their lives. They are hollow eyed, covered in mud, exhausted. A woman bends and plucks a swollen photo album from the saturated ground. She holds the book by one corner, and water pours out of its pages.

We are headed to the Hindman Settlement School, where both my husband, Jason Kyle Howard, and I often teach in the Appalachian Writers' Workshop, and where we first got to know each other. The school has been serving Appalachia for 120 years now, first as a boarding school that brought a classical education to an underserved population and now as a center for literacy, foodways, the literary arts, and more. The workshop, which has been a foundational literary gathering in the region since 1977, was in its fourth night when the historic flooding hit in the early morning hours of July 28. Some of its participants barely escaped before the flash flood got up to four feet deep inside their rooms. Others fled their apartments, which are built into the side of the mountain, hanging over Troublesome Creek, because there were fears of mudslides. They all gathered in one cottage, spilling out onto its porch, and watched the water rise. By morning they'd find that the offices, many of the classrooms, other buildings, and a lot of the campus grounds had been devastated. Many of their cars were carried away.

Jason and I are driving to the settlement school in hopes of helping. There are donation centers and cleanup crews set up in many places along our journey, but we are aiming for Hindman not only because it has been the hardest hit—seventeen of the dead so far are from around the town, including four siblings aged one to eight—but also because the school and the community are heartstrings of ours.

Once we arrive, we find that the settlement is a beehive. Volunteers unload cases of water from an eighteen-wheeler. A woman scrubs at a row of Blackstone grills where meals have just been prepared. A father and his two daughters carry plates of food up to the cottage where several families who lost their homes are now staying. The porch there is crowded with people watching dogs and children play together in the yard. All of them are strangely quiet, even the children who are swinging with their legs arching high in the air. The shock is mostly present in their eyes and in the way they move, stiff and somehow ghostlike. Other folks are unburdening their cars of diapers, detergent, paper towels, and canned goods they've brought to share.

The staff members of the settlement school, many of whom have suffered tremendous losses themselves, take action. The executive director, Will Anderson, helps unload the hundreds of cases of bottled water while overseeing the school's massive operation. The night of the flood, programming director Josh Mullins was in the offices trying to save what he could when the floodwaters tore off the heavy doors and rose above his waist before he escaped, copiers and desks and everything else churning around him. Today he is out on a run for more supplies. Rita Ritchie, the office manager, and Teresa Ramey, the bookkeeper, are working on the payroll; staff will likely need their paychecks now more than ever. Many others are at work all over campus.

The first person we see is Sarah Kate Morgan, the director of traditional arts. She is a phenomenal musician and a force who is always in motion. Since there is no functioning water at this time, she is unloading buckets that she will fill with water to flush the toilets of the people staying on campus. A day later, the maintenance foreman, Moses Owens, will figure out that a century-old well still works, providing the only running water for miles around. When I tell Sarah Kate we are here to help, she immediately fires off a list of things to do: we can unload or organize supplies, pull damaged furniture out of the ruined offices, help save the archives, and help prepare food or serve it to the community. Despite its own myriad problems caused by the flood, the settlement school has decided to open its doors and serve the region just as it always has. About forty people and twelve dogs are being housed there. The school is feeding and offering supplies to anyone who comes seeking help.

Usually disasters around here are ignored by the media or mentioned and lost in the next news cycle, but this flood has gained international attention. Two camera crews are present when we arrive in Hindman. While at this point they are mostly focused on the destruction, the help being provided at the settlement school is indicative of the general attitude of the people in the region. Nobody is waiting for anyone else to come do the work. The people have armed themselves with rakes, hoes, and shovels. They're moving debris and directing traffic alongside officials. Pentecostal church groups and organizations like Queer Kentucky are working together to organize collection centers and deliver supplies. A group of Mennonites from the area are set up on the side of the highway, dishing out fried chicken and mashed potatoes. Kristin Smith and Melissa Bond, married farmers who run the Wrigley Tap Room and Eatery in nearby Corbin, moved their services into the heart of the disaster area to cook for those in need. Groups such as eKy Mutual Aid, Appalachians for Appalachia, Foundation for Appalachian Kentucky, Appalshop, Pikeville Pride, and many others are on the front lines to help their own.

Melissa Helton manages community programs for the settlement. She is calm and decisive, and while focused on helping those most in need, she is also clear that one of the most pressing concerns is saving what can be salvaged of the school's archives. Most of them are wet and covered in the mud that has invaded everything. I am reminded of the woman pulling her family photo album from the muck. I am reminded of how my own mother grabbed only our photo albums and me when a flash flood took everything our young family owned decades ago.

Jason and I go to work after a quick tutorial by Melissa, who has been taught a few tricks by trained archivists who have volunteered their time over the last couple of days. Many of the items are beyond saving. Some of the photographs can be dried easily. Others must be dunked in water to remove the coating of mud, lightly dabbed, and hung or laid to dry.

The archives hold not only the history of the settlement school but also that of the region itself. This loss is particularly devastating in part because Appalshop, a media, arts, and education center in nearby Whitesburg, was badly hit as well and houses even more archives of the region. At Hindman there are thousands of items detailing the daily lives of Appalachian people over the last century, showing everything from folks outfitted in suits or full skirts while enjoying an early 1900s hike to children learning in computer labs just last year. There are recordings of legends like Jean Ritchie performing onstage and 1960s children participating around a maypole. I find a whole cache of postcards that were sent to the school's staff by James Still (author of one of Appalachian literature's foundational texts, *River of Earth*) during his travels across Central and South America.

There are hundreds of pages of letters written by locals and the schoolteachers who came to the mountains from all over the country to participate in a social educational experience that changed the region for the better and old brochures from the writers' workshop with a photo of beloved writer Lee Smith on the cover. Smith has been a longtime supporter of Hindman Settlement School and is the person who first sent me there. Dolls have survived that were handmade by Verna Mae Slone, a local woman whose 1979 memoir, *What My Heart Wants to Tell*, became a surprise bestseller. I even find pictures of myself during my long association with the school, including my first author photo, taken on the creek's banks twenty-two years ago.

The lives are the most important and disturbing losses, of course. The day we were in Hindman, another body was found not far downstream from us. As of this writing, thirty-seven people are dead from the flooding, and Governor Andy Beshear says we should expect that death toll to rise. To be in eastern Kentucky right now is to shed tears of sorrow, pride, anger, and hope, sometimes simultaneously. So many are in the depths of the worst kinds of grief. You can feel it when you're driving those winding mountain roads, when you see people living in tents or shelters, when you see children trying to play when they know their family has lost everything they own. Seeing them, I can't help but blame the coal companies who disrupted so much natural drainage, caused widespread erosion, and did more damage that directly contributes to these disasters. I can't help but blame the legislators who refuse to pass climate initiatives.

Today a heat wave has moved into the region again. The sun is beating down on the backs of people working hard to save what they can, to recover bodies, to clean out the seemingly infinite mud and muck. But the sun is also helping to dry up the saturated ground as well. The sun, though hot, is still shining for those left behind. And the people keep working.

Food, cleaning supplies, and other donations for the community in the Great Hall. © Tyler Barrett

Wes Browne
From the Road

THE SECOND TIME WE CARAVANNED INTO THE FLOOD-RAVAGED areas of eastern Kentucky with our pizza trailer, I was heartened by how much better things looked from the road. But it made me nervous just the same. When things get to looking OK from the road, that's when a lot of folks forget what's beyond the tree line and past the next hill.

I first experienced the shock of the 2022 flood through the lens of friends on the campus of Hindman Settlement School for the annual Appalachian Writers' Workshop. Aside from my own home and the one where my wife grew up, I've laid my head down on that campus more than any place in Appalachia. I wasn't there the night of the flood, but I saw the devastation through the pictures and stories of friends who were there that night and in the days that followed.

Once I knew everyone at the settlement school had survived, my mind, like many others, turned to what we could do. "Don't rush in here if you don't know what you're doing." That was the message to would-be Good Samaritans who wanted to help but weren't quite sure how. I can swing a hammer, but I was far from the kind of laborer they needed.

Instead of rushing in, my partners and I set to what we were better equipped to do. We raised money, collected supplies, and made plans to go in with our Apollo Pizza trailer and get back out without taxing any resources. Since my closest connection was to the settlement school, I counted on their staff for guidance.

Our customers donated, and we matched the amount. Some of our suppliers cut us price breaks so we could stretch the money. We also collected diapers, wipes, tampons, and other personal hygiene products, which were in short supply. After watching the weather and abandoning a couple of plans to go in earlier, we finally left for Hindman the morning of August 10, twelve days after the flood.

My sixteen-year-old son and I drove a refrigerated van full of supplies. Our maintenance man and his fiancé hauled the food trailer. One of our managers drove a third vehicle packed with the personal hygiene products.

The first hour was much like every other trip I had ever made to Hindman. Hills and highways and small towns. That was until we got to Jackson. Jackson's business district has never been what anyone would describe as glamorous. It's characterized by small strip centers, fast food, and gas stations. It's also built along the North Fork of the Kentucky River. The first signs of the devastation appeared in the form of wreckage piled in parking lots alongside backhoes, dump trucks, and other machinery. Some of the equipment was at work piling and filling trucks as we passed by.

It was around Jackson where we began to see foundations without houses. House trailers on their sides or their roofs, with every remaining belonging piled in the yard. Cars and trucks and trailers down in the river itself. Debris scattered fifty feet up in trees, looking like it had all been thrown in a food processor and spilled out into the river, along the road, everywhere.

My son, as sixteen-year-olds do, was trying to get a phone signal. "Look," I said, tapping him on the chest and pointing across his body. "All those concrete pads were people's houses. Two weeks ago they were living in them, and they're gone. Some people barely got out. Some people didn't get out. You understand that? What if that was our house?"

He said he understood. We repeated this countless times as I pointed to variations of the same devastation, and I made him look. "I understand," he said. "It's bad," he said. He'd go back to his phone. But sometimes I'd watch him from the side of my eye, and he'd take it in for long stretches when I wasn't prodding him.

The only place I'd ever been to compare it to was the Lower Ninth Ward after Hurricane Katrina. The remnants of homes. Foundations and porch steps with no house to go into. That's what the flood was to eastern Kentucky. A hurricane raging its way down the river, leaving behind much the same result.

Our path through Hazard was mostly on high ground. Hazard got hit, but the area we saw looked a lot like always. The worst devastation in evidence was the mountaintop removal mining site visible from Kentucky Route 80 on the way out of town. That human-made catastrophe looked the same as ever. It also stuck out as an example of why the flooding was so bad.

Surface mining sites and the flooding had a strong correlation. Trace your finger on a map across the most exploited mining sites in eastern Kentucky, and you'll run it across many of the worst-hit communities. The mining robbed the area of porous topsoil that was never adequately replaced, leaving behind impermeable, flattened terrain. When rain hits those sites, it runs hard and fast. Combine those conditions with rising temperatures that lead to more severe weather, and this "act of God" has the fingerprints of humans all over it.

We reached the turnoff for Hindman on 80 and headed downhill toward the settlement school. The first piles of detritus from the flood sat on the high ground near the funeral home. We were running behind schedule, so there wasn't much time to reflect, but at the bottom of the hill there was no choice. Even after days of cleanup, every direction was rubble. Downtown still stood, but it had been hollowed out. And what it had been emptied of lay along the streets. A handwritten sign at the roadside asked that people keep their advice to themselves.

As we were crossing the Jethro Amburgey Bridge over the beloved but painfully well-named Troublesome Creek onto the settlement school campus, it occurred to me not for the first time that people can become fond of what sometimes hurts them.

The settlement school looked a bit like it does on the first day of the Appalachian Writers' Workshop: a hive of activity. The big difference was nobody was smiling. All faces this day were strained and resolute, unlike the low-grade euphoria seen when the workshop kicks off. And like everywhere else we had been, there was the rubble—the piles of what had been cleaned out but not cleaned up.

We were greeted warmly by some faces I knew and some I didn't, but we did what everyone else was doing. We set to work. There was a long line. We plugged in our big 220 cord but couldn't plug in the refrigerated van so just had to leave it running. We made do. I assembled the pizzas and put them in the oven, our manager cut them, our volunteer handed them out, and my son restocked our supplies as we went.

It didn't all go smoothly. Our rule was a limit of two pizzas per person. The first person in line was a Kentucky state trooper. He ordered twelve pizzas. Knowing how long that would take and how many hungry people were in line behind him, I admit I was frustrated. A young member of the settlement school staff whom I had just met smoothed things both ways. She explained to him that we couldn't give him that many but that there was more food inside he could take to his people. She told me that a lot of the troopers were struggling. That's who had been dealing with the dead, and they were exhausted. We made him four pizzas, and we moved the line.

We rolled out over two hundred pizzas in two hours. Our crew was upbeat but all business. My forty-eight-year-old back ached a bit, but it seemed like a petty complaint. All day we saw people who had lost everything, or close to it, and by and large their heads were high, and they thanked us. They managed a smile; they ate hot pizza. As one of the staffers told me, "You're not going to cure everything with pizza, but stuff like this helps keep people's spirits up."

After the settlement school, we made our way to tiny Garrett, Kentucky, in Floyd County, where the lone store/restaurant in town had been wiped out. I'd been there a few times before because it was operated by my wife's distant family. The store and town sit on low ground alongside railroad tracks. Folks were lined up down there waiting for us.

It was like everywhere else we'd been. The entrails of the place were on the outside. Worse, the mud that covered everything smelled of sewage. I warned our crew to sanitize early and often.

The store owner was feeling upbeat. A restoration company was there from Lexington to clean out and disinfect her store. "I'm hoping to be back open in three weeks," she told me.

"That's great," I said. What I didn't say was how uneasy it made me. Would they flood out again in a matter of months, years, or decades? I hoped hard it would be the latter.

If anything, the people in Garrett were even more grateful to see us than the ones in Hindman. The settlement school had gotten a lot of attention. The archives there had flooded, and a trove of Appalachian treasures were lost or in danger of being lost if they couldn't be restored. The food and supplies that had gone in there were far greater than what they'd seen in Garrett.

I heard someone complain that there were "junkies" in our line, suggesting they didn't deserve to be fed. All I could think to say was, "Hopefully everyone will get well." It seemed like a strange time to ask questions about who you should and shouldn't feed.

Before leaving Garrett, we made thirty pizzas for a remote nursing home that felt forgotten. The folks from the store volunteered to run them over there so we could move on to the Knott County Sportsplex, our last stop of the day.

As we chugged back up 80, the sky turned charcoal, and the trees began to sway. The woman who met us there spoke plainly. "Everybody's leaving. People aren't taking chances with the rain." She looked at the periphery, where the skies had turned. "You can stay and try if you want, but you don't want to be here if we flood again."

With that warning and the lack of people, we didn't even unload. We'd bided our time because of rain forecasts and had been lucky so far. We weren't looking to push it.

We still had supplies to make more pizzas. After a few weeks of stops and starts, and after finding that the Sportsplex operation had largely been wound down, we made our way back into the hills of Whitesburg in Letcher County on September 29. We planned to set up at a community center called CANE Kitchen, where more hungry people were waiting. While in Whitesburg, I hoped to look in on Appalshop, another Appalachian touchstone with archives that had been hard hit.

Most of the route to Whitesburg is the same as to Hindman. My son was back in school, so this time I drove a trans college student who works at Apollo and had volunteered to help. I was impressed by his enthusiasm to give, and moreover by his enthusiasm to give in a setting where he didn't know how he would be received—but then again, that's everywhere.

The difference in the drive wasn't just my traveling companion but also the fact that instead of telling him how bad everything looked, I found myself pointing out how much better it all seemed. In a month and a half, most of the debris was gone. Houses were being replaced and rebuilt. Life appeared to be going back to normal. Appalachia is resilient, and sometimes it's deceiving. Beyond what we could see were abandoned houses and displaced families. Some that were never coming back. The FEMA money wasn't enough to rebuild, and in some cases the homesites had been rendered unfit for it.

There are some clear-eyed projects intended to acquire land on higher ground in hard-hit communities. To build homes away from the rivers and creeks that overran. For it to work, people will have to give up homeplaces going back generations. While some folks welcome a new start, not everyone does. It's both understandable and concerning. Although I wanted to be happy for the people rebuilding at the rivers' edges, I was fearful for them.

There's so much still to be done that the need feels almost endless, and people are already forgetting. Moving on to the next disaster before this work is even done. In February 2023, as if to make sure we didn't forget, Hindman and other communities flooded again—not like before, but enough to remind everyone what the water can do.

When the next big flood comes, will eastern Kentucky be rebuilt from the last one? And how much of what's been rebuilt will be torn down again? I fear we'll find out sooner than we'd like. It's not the view from the road that matters. It's the view from the homes. The outlook going forward is an uneasy one, and that may not change for a long time. If it ever does.

Amanda Jo Slone
Salvage

(verb) *to save from destruction, damage, ruin, or loss.*

I return to Hindman one week after the flood. Seven days have done little to ease the anxiety of that night but have transformed everything I see beyond my windshield. In some places, thick mud blurs yellow lines. In others, full lanes are reduced to half, and shoulders droop. My father drove me to Hindman for my first Appalachian Writers' Workshop in 2007. I could have taken myself down the familiar stretch of Route 80, but the ride with him was filled with singing, with Dad pointing out the places he remembered from his time in Knott County coal mines. He hasn't been behind the wheel of a car in more than three years, but as I drive by piles of things that a week ago were not garbage, I wish my dad was in the driver's seat.

I sit cross-legged and surrounded by folders on the wooden floor. I snap purple gloves on my fingers and listen to a brief orientation. I am not qualified to work with archives, but the directions seem simple: Peel the photos from their original pages and place them on dry paper. If the descriptors are legible, write them on the top, and clip the new pages on a clothesline to dry. I am surprised by how many of the photos are still clear, how durable older images are. I have no talent for placing things in time, but I know I will not recognize anyone in the black and whites. Some faces smile beneath the mud; some are stern and stiff. There are a few plump and bonneted babies misplaced among their elders. I try to work quickly, with the sweet smell of rot an offensive comfort above my lip. I swallow it down to keep myself from choking on the gravity of the history I am handling, the messy responsibility of someone else's memories. I repeat the instructions to myself like a mantra: Peel, stick, clip. Try to apply the old ordering system to the new.

To make a photo is to freeze a moment out of context, to hold it immutable, captive and timeless, subject to interpretation. In some of the soggy moments from the archives, I recognize landscapes and buildings, places I've felt connection on the settlement school campus. Places I've made memories, made photos of my own. At home there are trunks spilling over with photos, albums grown dusty on shelves. I think of the moments contained inside, the ones I felt were important enough to shoot and print—moments I realize now I hoped would carry a story beyond me.

In one old family album, my father is young, plucked from eastern Kentucky by a recruiter who likely promised him a life beyond the coal mine he'd been crawling in when the roof fell and killed his own father. He was barely eighteen and didn't even tell his mother he was gone until the first chance he got to write from basic training. He remembers filing off the bus, voices yelling for him to place his feet in yellow footprints painted on the concrete, someone calling him Hillbilly, the nickname sticking. He remembers Lynn Anderson crooning from a loudspeaker, claiming no promises of rose gardens, acknowledging that along with the sunshine, there's gotta be a little rain sometime. He remembers those distant moments even on the days he does not recognize my mother.

We received the first letter from the commandant of the Marine Corps in 2008, thirty-six years after my father left Camp Lejeune and twenty-six years after they discovered chemicals contaminating the drinking water. This information, of course, was diluted with patriotic platitudes, just enough thanks for his service to help us swallow the news. It took a decade for the diagnosis, and though every letter, pamphlet, and card that came in the mail warned us of the connection between the poisoned water and Parkinson's, we were surprised when it came. We were surprised when his muscles weakened, when the hinges of his hips loosened, and when he began wading through a murky reality.

TROUBLESOME RISING

To make a photo is to preserve a feeling, to love it into a shape that can be carried through all the things that happen in life. Remembering. Forgetting. Creeks that jump their banks. Water that is poisoned. The task is heavy. I am not qualified. I pull up my purple gloves and reach for another folder. I repeat the mantra: Peel, stick, clip. Try to apply the old ordering system to the new. I salvage what I can.

Crisis therapy dogs greet and comfort community members. © Tyler Barrett

Scott Honeycutt
To Ask

Where have they gone—those rivers that drowned themselves
dry? Where have the crossings, nightfalls, and a fat buck
that stood statue-still in the fast lane gone?
Where are the vultures, the hawks?
Where the roadkills, the entrails and coyotes, the blood patch of dogs?
Where, you?
What swept up those rails and bridges?
Who gathered the maps and washed clean the miles between?
Why born to loss, a wind-tossed and wind-slipped silence?

August icelike in those dark alleys of Kentucky—
Tunnels, voiceless snow, like silence after
a climber loses his grip on ice and falls,
far distances swallowing his voice—What and where?

The doves will be back, for sure,
constructing nests under our porch eaves,
but will we still be us? And what us will await their return?

Amy Le Ann Richardson

It Is Still Here, the Magic

1. Before

When it was all sunshine.

When legends of those before whispered to us
where our roots run deep and stories are stored,
stacked in layers like rocks of mountains
wrapping around us.

When we stayed up too late listening
to teens trade tales at Ironwood,
make connections, grow a community.
Our biggest worry, their experience

in this place we cherish.

When we led them with pages and ink and
a techno-contra-dance myth of their own to pass on,
and walked them through the greenhouse,
blessing the inaugural summer with sweet promises

and dulcimers rang out in the Great Hall
enthusiastic bum ditties galore.

Amy Le Ann Richardson

When we shared our own sagas in circles
on porches at the Appalachian Writers' Workshop,
words echoing amid a chorus of bugs and owls
into possum homes tucked just beyond our light.

When we stayed up too late, slept too little,
crammed words between with coffee and brown butter cookies,
our energy spent, but hearts and minds soaring.

When we walked up the hill after trivia that night,
commenting on thunder and dodging raindrops,
none of us knew those were the last moments before
it all changed in this place.

2. In the Beginning

It rained all day, then stormed well into dark,
lightning flashed through the windows as I tossed and turned,

pulling up the weather app on my phone one more time,
alert after alert vibrating, jarring me, each one increasing the inches of rain,

increasing warning severity, increasing anxiety, driving thoughts
to my children counties away, hopefully safe, asleep, unaware.

Still in the hazy space between slumber, which never fully came, and awake,
frantic knocking jerked me from bed, electric buzzing in the air,

I slipped on shoes, opened the door. Flashlight strobes punctuating
the night between lightning strikes, my stomach rolling in rhythm,

my mind fighting to absorb words, information, what was happening,
this unreal present washed in the sick glow of emergency lights

haunting spaces where I'd laughed only hours before.
I stood with a gut of lead in this place I adore.

3. Then We Saw the Water

I placed both palms on the railing, holding myself up,
where we huddled on Stucky's porch, staring,
rain still pouring, the creek roaring,

I pushed slick water with both hands back and forth
along the length I could reach, trying to ground myself,
the only tactile sensations surrounding me,
all tied to the flood my brain wasn't comprehending.

I heard exclamations, expletives, ongoing chatter
as each of us saw the water,
only visible in flashes of lightning,
rising as we watched,
nowhere to go but roadcuts through mined mountains,
nowhere to go but up, toward us.

My palms pressed tightly in place as
gasoline permeated each breath,
I heard others gasping, my own lungs tight,
I turned back inside, pacing between
panic on the porch and calm on the couches.

I sat in a rocking chair,
burning anxiety in short pulses instead of paces,
listening to escape plans and reassurances,
watching panic overtake my friends like
the flood rising outside,
already swallowing cars and buildings
we'd believed safe.

Hours of dark, I rocked,
glancing up through the skylight,
watching for any sign of day,
seeing only lightning for so long.

We wandered, switched positions and places,
waiting for news from our unaccounted-for friends,
from town, from the world,
missing constant connection of Wi-Fi and cell service,

but there was no reaching us in this place.

4. In Daylight, We Left

The hazy gray air and roar of rushing water
engulfed my senses as I walked toward Mr. Still's grave
when light touched treetops.

A stream pouring down the hill
covered my feet with sharpness like ice,
I looked beyond the chapel to Main Street,
my chest heaving, breath sucked out into the mist as

I watched muddy water roll over it all.

I joined a group gathered walking toward town.
Water receded enough to cross the bridge by Mi-Dee Mart
where we paused on the corner of Main Street,
water still gushing, pushing pieces of houses and
bits of lives down the middle of the road, and
people gathered gazing over the waterline

looking for answers, looking for fragments of their
lives,
looking for lost loved ones,
finding only their living room carpet and quilts.

I'd witnessed floods before, seen disaster,
but nothing like this, coming while we slept,
nine inches in an hour with creeks at capacity,
taking anything in its path.

TROUBLESOME RISING

Hearts sunk to the bottom of Troublesome,
tethered to the town we love,
we walked back, loaded up, and arranged carpools to leave.

Getting out of the way, the best help we could offer Hindman right then.

Arriving to a whole family, untouched house, and warm bed felt so unfair.

How I could witness such hurt, such loss, and
have everything the same as I left it at home?

How lucky I was, even with these still-stark shocks
down my spine every time it rains,

pulling my mind back into this place, underwater.

5. Pieces Everywhere

Restless at home, I drove back the next day
with everything I could afford to buy at the Dollar Store

to pull our history from muck jammed into every crevice
covering books and journals and pictures,
irreplaceable items submerged.

I dug through a tote bag of keys
caked with moist earth to unlock archive drawers.
I pushed wheelbarrow loads of files uphill,
peeled photos from frames, gently placing them between
moisture-wicking papers in hopes they'd dry.

I witnessed love between strangers.
All of us pulled to this place
eager to preserve what we could.

So many hands passing pages
and instruments and stories and disks and dolls,

Amy Le Ann Richardson

hanging them to dry on makeshift clotheslines
across the stage of numerous readings and
speeches and dinners and dances,
mud streaking the floor and our faces,

a time line stretched before us,
the tapestry woven by so many folks.

Community streaming in for weeks clearing debris,
serving food, sorting supplies, mopping mire,
rescuing legacies left on those creek banks,

hope rising higher in this place than water ever could.

6. After

Recovery never ends, but now,
it feels almost normal.

Reality warped like it was left in the sun,

parts exactly the same in those creaky dining chairs
and the walk uphill where I get out of breath just at the top,
the green couch in Stucky where I sink almost to the floor,

parts beyond repair where mud sat untouched in leaning buildings,
displaced offices all tucked together in the Gathering Place,
road signs and toys still stitched into the creek bank

where I walk, bending to touch buried books,
run my fingers across blurred words and frayed spines,

where I look up to those same blue skies,
hills kissing edges to hide the horizon.

Where instead of running, I want to return,
bring my children, and show them their heritage.

TROUBLESOME RISING

Show them what resilience looks like. Fortitude.
In this place. In its people.

Give them hope to hold in this uncertainty
amid our shifting climate and riots and pain,
show them how we survive. Together.

And it is still here, the magic.
The pull to visit, to gather,
to croon tales into the night,

to see what else we need to learn and
share sagas in circles on porches

in this place.

Richard Hague
After

> *Any community is dependent on place, even in our web-wide, pixelated, Zoomified world. To witness the flood waters rise to sweep away so many lives and homes and cultural artifacts—that damage to this place felt like damage to my own body, a physical injury. And months later, when we were able to gather again on this sacred ground, that too felt like healing, like the body finding its way again back to health. Scar tissue is stronger than skin.*
> —Jim Minick

IN THE LOWER PARKING LOT, A FEW YARDS FROM THE FORKS OF Troublesome Creek, four white vans are jammed in a tight line together, so closely you couldn't open any of their doors except the outermost one on either side. Debris of all kinds—branches, limbs, plastic, shreds of cloth—is lodged underneath them. At first I think the vans must belong to volunteers come to help in the cleanup, but then I see that the last time they had moved was when the floodwaters smashed them all together and began to fill them with an inhuman miscellany of flood-stuff. Just a few feet beyond, at the end of the bridge into the lot, a small compact car has been pushed to the side. As if growing from the front seat, the trunk of a small tree, branches and bark torn off, juts through from where the windshield had been: a monument to apocalypse.

I am with several writers who had come to Knott County weeks before to help clean up after the deadly late July 2022 flood and have now returned. Some, like Pam Campbell, Roberta Schultz, and Jim Minick, had actually been attending the Appalachian Writers' Workshop when the flood struck. It is October, time for the meeting of SAWC, the

Southern Appalachian Writers Cooperative. For decades, we held our annual meeting at the Highlander Center in New Market, Tennessee, but it remains closed to groups due to the pandemic and to ongoing concerns after a 2019 act of arson—perhaps a hate crime—took out their office and destroyed many of their records. The Hindman Settlement School, equally important to all of us Appalachian writers, has agreed to host us as their first official group event since the flood. We are grateful, pleased to be able to attend some sort of normalcy after a summer and early fall of funerals and ceaseless cleanup. The Mullins Center Great Hall is filled with boxes of salvaged books, files, and carefully laid out photos and documents drying on tables and the floor.

Grief is still in the air. In the room where for decades many of us, both as attendees and staff at the weeklong Appalachian Writers' Workshop, listened to the greatest writers of the region talk and lecture and read and sing, the mood is subdued. And though the meals from the newly recovered kitchen are as nourishing and tasty as ever, we can't help but feel the somber nature of this year's gathering. It's as if ghosts are everywhere: not only those of the past writers we studied with—James Still, Jim Wayne Miller, Harriette Arnow, and the founder of the workshop, Albert Stewart, to mention but a few—but also of those killed in the flood here in Hindman and all up and down the hollers of the surrounding counties, forty-four souls all told.

Floods are frequent in Appalachia, but this year's was remarkable for its intensity, its death toll, the cultural devastation it wreaked, and the long-lasting cleanup ahead. The force of the water and extent of the destruction was, as more than one eyewitness mourned, "unimaginable." The black iron of the iconic campus footbridge across one of the forks remains bent; we are left to imagine what huge tree trunk or small car barreling downstream must have bashed it in. (Jim Minick heard it was a huge propane tank.)

The damage remains clear. In the manner of one of Pauletta Hansel's collective poems, condensed and lineated from notebooks and recollections of some SAWC members who were there during the flood or for the cleanup afterward, come these witnessings.

Roberta Schultz, there during the flood, recalls the ominous night and the morning after:

Richard Hague

Rain started in earnest
after midnight
pounding heavy
on the roof

alerts changed to
Severe Emergency
Life Threatening

Participants evacuated
moving cars
up the road
toward our cabin

gas tanks ruptured
into water supply

circuitous routes
passing damage
and large debris
in every county
leading out
east and west

home to learn
of rising
deaths

    Judy Jenks, during the early August cleanup, observes painfully the luthier's workshop downtown:

Muddy wood,
buckled floors,
flood line
five feet
up the walls

Pam Campbell, also there during the flood, returned during the August cleanup and dreams of recovery:

While they build
beautiful mandolins,
guitars, and dulcimers,
. . . rebuilding
a life full of purpose
and artistic creation

Judy Jenks again, on the aftereffects:

The red car is
still upside down
as if it belongs
there now.
One of those
weird art installations
—symbolism—
as if a car upside
down in a creek
is the natural state
of the earth.

Scott Goebel, volunteering at Hindman two weeks after the flood, also goes on to Whitesburg to help at the damaged Appalshop; he writes on returning from an attempt to check on a missing person (later found to be safe):

The stretch from
Lost Creek toward Quicksand
was the worst.
I didn't realize
until I got back
that Troublesome
was only one
of four big creeks

that dumped
into the North Fork
near Lost Mountain.

I didn't get a chance to look,
but I suspect
that it was made worse
because they started stripping
Lost Mountain again
four years ago . . .

Tents everywhere.

Tarps strung up
between trees.

 Because so many SAWC members have attended the writers' workshop for decades, the damage to town and the campus itself is particularly painful. It is painful also because we recall that many of SAWC's earliest members, including the late Jim Webb and Peggy Dotson, grew up in areas hit hardest by the flood. Pauletta Hansel, another native of the region, writes, "All of us have benefitted from Hindman's work—from the strong sense of community and literary excellence it promotes. Honestly, I can't imagine our gathering anywhere but at Hindman in 2022 and am so grateful they wanted to have us. I felt helpless after the floods, and we all did, those on the ground and those away . . . survivors' guilt surely, but for me too a sense of dislocation, homelessness. Being a tiny part of flood relief while hosted by Hindman in August and coming back in October for SAWC helped me put some of my missing pieces back."

 The gathering after the flood also helped SAWC regrow its membership postquarantine, reassuring SAWC co-coordinators Nicole Rahe and Owen Cramer. The location in Central Appalachia drew new and returning writers from places like Kentucky, West Virginia, Ohio, and southwest Virginia. More new members gathered than at any time during the last decade. "The Settlement School," Pauletta says, "was the perfect place to ground ourselves once again in the history of SAWC and its place in the Appalachian Writers movement." Veteran SAWC member Dana Wildsmith, one of the organizers of the cleanup but who couldn't be there, agrees: "When your world turns on you while you're sleeping, you turn to family for help and comfort. Appalachian writers are cousins of the heart, and Hindman Settlement School is that heart. We have spent summer after summer in reunion at the Appalachian Writers' Workshop, leaving that week renewed in our connectedness to each other and resolved to work for the good of the region through our words."

One of the most hallowed of Kentucky writers, Wendell Berry, frequent visitor and speaker at the Appalachian Writers' Workshop, is well known and beloved by members of SAWC. His own work provides cultural context for the renewal and coherence of the group and the settlement school after the flood. Wendell Berry reminds us that a healthy community relies on and requires its own powers of support. No government can solely do it; primary are the communal efforts of the people.

Along with the members of SAWC and the staffs at Highlander, Appalshop, and Hindman Settlement School—and the active writing communities and artistic and intellectual neighborhoods they mutually support—we offer to Mr. Berry's words a collective "Amen."

Southern Appalachian Writers Cooperative was the first retreat hosted postflood. © Melissa Helton

Jim Minick
After the Flood, After the Tornado, and Before the Next

1. A New Understanding

After the rain whipped us into the night corners of porches where we watched in silence; and after a few hours of worried sleep where I dreamed the cabin slid down the mountainside; and after waking in the early light to find the valley unfamiliar, with shallow Troublesome Creek now a twenty-foot-deep river filling the classroom where just yesterday we had talked about poetry and wonder; and after walking to find the other writers huddled on porches stunned and staring at the family of frantic ducks washed up on a lonely spot of grass;

and after the waters receded to reveal cars jostled together or upside down in creeks; and after trying to rescue books and valuables from these cars for friends, finding instead console cupholders filled with mud and books swollen and dark;

and after driving away, feeling like abandoners and yet knowing that if we stayed on as the guests we were, we'd only be more weight on an already weighted-down place; and after slowly steering around downed trees and mudslides and washed-out bridges and buckled roads and flattened gardens and foundations suddenly empty where houses once stood; and after making it to Whitesburg to find larger bridges covered in still too much water, the line of tractor trailers idling in the sun; and after finding a way over Pine Mountain into Harlan and eventually home to Virginia, where I kept looking for pink kiddie pools high in trees and windows slimed with mud;

days after all of this, I read this comparison: "Water moving at over 6 mph can exert the same pressure of air moving at EF5 tornado wind speeds—200 mph or higher,"

and I finally understood

that I had been writing this story already for over a decade,

just set in a different state, Kansas instead of Kentucky,

and a different time, 1955 instead of 2022;

and that no matter how much I tried to imagine my way back into the horror of living through a tornado, that experience was beyond my grasp, I hadn't lived it, and yet, now, in a way, I had;

and that no matter how futile our attempts to write these stories of disaster and recovery, these stories of resilience and community—no matter the struggle to find every needed word—

I understood these words might be all that survive,

and so, after all of this, we'd better get them right.

2. The Facts

On May 25, 1955, an F5 tornado struck Udall, Kansas, at 10:35 p.m. There was no warning. In roughly three minutes, it destroyed most of the buildings, toppled the water tower, and killed eighty-two people. The Udall tornado was—and still is—the worst in the history of Kansas.

In the early morning of July 28, 2022, a massive storm stalled over eastern Kentucky, where it dumped over eight inches of rain in only a few hours. There were warnings, but many people, including me, just wanted to sleep, couldn't fathom the danger, didn't want to remember how centuries of mining had compromised these mountains, and so we stayed in bed. Water swelled to break dams, crush bridges, roll houses away, and kill forty-four people. The eastern Kentucky flood was called a thousand-year event, one of the worst in the history of the state.

3. The Backstory

Without Warning: The Tornado of Udall, Kansas is a book I've been working on for over a decade. I've shaped a narrative of this event based largely on stories from survivors whom I've come to call friends.

Through it all, I've had to wrestle with this conflict: this is not really my story—I didn't survive the tornado, I've never lived in Kansas, and I wasn't even born in 1955—and yet, to write this book, I've had to imagine my way back into that horrible night and the lives of these survivors and all they suffered. I've had to make the Udall tornado story my own.

Living through the Kentucky flood proved to me how impossible that really is, how impossible, really, the acts of writing and reading are; we can never fully relive an event, never fully recreate a time and place and the rawness of emotion, *even if we lived through it*. And yet, as writers, as artists, as creative people, we know that these tasks of reimagining the past *and* the future are essential and impossibly necessary. Our acts of creativity, our arts, define us and create the narratives we live by, and as the climate crisis has revealed, if we live by bad stories, we are bound to die by them as well, unless we can imagine better stories. As creative people, we have an essential role in shaping narratives that shape the people and places we call home.

4. Earl Toots Rowe, Mayor of Udall, Creator of Story

In 1955, Earl Rowe, whom everyone called Toots, was the just-elected mayor of Udall. On the night of May 25, he was about to head to the oil rig a few miles away where he worked the night shift. Before he could leave, the storm struck. Toots and his family didn't have time to escape to their storm shelter, so they crawled under their dining room table—Toots; his wife, Lola; and their three children. The tornado blew out the windows, and then one wall of the house collapsed onto the family. A brick hit Toots on the head and knocked him unconscious for a few minutes, but somehow, he woke to crawl out from under the wall, and somehow, he lifted that wall enough for his wife and children to crawl out.

After Toots got his one injured child to a hospital and the rest of his family found shelter in a nearby farm, Toots returned to help with the search and rescue. For the rest of that night and all the next day, he and others dug through the rain and rubble to carry neighbors, friends, and kin—some dead, some alive—to waiting ambulances.

One newspaper reporter asked Toots how he felt. He replied, "I still can't believe that it's all happened—that so many of my friends and neighbors are dead or injured, that we don't have a town to live in, that a number of them probably still lie buried beneath the rubble." He paused.

"My house just floated away. I don't know where it is." He looked around. "There's nothing left." He repeated, "There's nothing left."

When darkness fell the day after the tornado, Toots was so exhausted he had trouble thinking. He had been awake for almost twenty-four hours, and not just awake but frantically scrambling over rubble searching for anyone still missing.

Now, though, with the horizon a milky blue, it was time to rest. He shouted to others to call it a day, to go home—wherever that home might now be—and to rest for the night. They had rescued the living, sent them to hospitals. But some people were still missing, and they expected to find more bodies under the rubble the next day.

In his tiredness, Toots's wounds throbbed. The hole in his leg—a puncture the size of a nail—seeped and stained his pants. The back of his head ached where a brick had knocked him out. But he couldn't sleep, not yet; he had to stay focused.

The mayor stood in the middle of the remains of his town. A breeze rocked a tin sheet of roofing somewhere out in the rubble, but otherwise the place was silent and empty. No trees caught the wind. No leaves blocked the night sky. The stars blinked on, like every night, when someone flipped that switch, but before him, no houselights or streetlights. He felt a sudden loneliness as these last hours—the whole enormity of it all—settled in.

Two men joined Toots, officials from the state and federal governments. Toots told them what he could about the dead and injured and the number of buildings destroyed.

All the while, though, he knew their ultimate question.

After some silence, one of them asked it. "Do you want to rebuild?"

"Yes, I want to rebuild. We have to rebuild."

The two men glanced at Toots Rowe, tried to see his face in the darkness.

"But there're no people," one of them said. "You can't have a town without people."

"People will come back. Just give us the money for schools and utilities and churches. You'll see."

"But why do you want utilities? Or schools? There's nobody here."

"They'll come back," Toots said. "We have a new water and septic system, so we just need the electric lines restored. We can reuse the schools' foundations. Same with the churches. You help us, and the people will return."

Even in the emptiness, Toots felt the spirit of so many. He couldn't walk away and bury it all.

"You give us those things, and the people will come back."

The government men agreed to help.

The next day, while searching through the rubble, Toots met Udall city clerk John Arbuckle. Arbuckle had been paralyzed in a swimming accident as a teenager, so he used crutches to get around. Arbuckle's sister had come to take John to her house in another town, but as they drove away, John saw Toots and rolled down his window. They greeted each other and talked about how they had survived, and then John asked Toots what he was going to do.

"This is my home, and I plan to live here," Toots said, no hesitation. "I'm going to rebuild and stay."

"Well, if you are, then I am too," Arbuckle said. "I'm staying too." He got out of the car, using his crutches to maneuver over the debris. He couldn't help with the search, but he could still be a recordkeeper, a source of information on the dead and injured. Toots found him a place to sit, and the city clerk went back to work.

Another friend, Wayne Keely, met Toots during the search. Keely was the city marshal—he knew and loved this town. The two men shook hands, and Keely told of how he had heard the roar and yelled for everyone in his house to get into the cellar, where he closed the door just before the tornado hit.

Keely's wife asked if it was the train she heard. Wayne shook his head and said the train had already traveled through; it was a tornado. As it passed over, Keely's daughter feared the cellar would collapse, it shook so much. Dust filled the small space, making it hard to breathe.

Then the roaring subsided, and Keely opened the cellar door, shoving away debris.

"What's it look like?" his wife asked.

"I don't think there's ten people left alive," he said. In the lightning flashes, he saw no houses, few trees. He could see all the way to the other side of town, twelve blocks away. Right beside him, the remains of his house had crushed his police car. And in his front yard, the tornado had hung the frame of a pickup twenty feet high in the fork of what remained of a tree. The truck's body, as one reporter later described it, "was wadded up like tinfoil."

After the storm, Toots asked Keely, "What are you going to do?"

Keely said, "I'm gonna build somewhere else."

"The hell you are," Toots answered. "You're gonna build right back here."

Keely eventually agreed.

Many, like Arbuckle and Keely, heard Toots and stayed on because of him. They picked up his words—*I'm going to rebuild*—and made a refrain of hope, resilience, and determination.

5. The Stories We Create

Somehow Toots Rowe knew it was his job to create his town's story of survival. He had to tell the world about the horrible loss, to put words to the pain. And he had to tell Wayne Keely: *To hell you're leaving. I'm rebuilding, and you're still going to be my neighbor.*

What stories are people who experienced the Kentucky flood creating about surviving, about rebuilding? And what stories are we all creating now to carry us forward through the great upheaval called the climate crisis? One version, the easy one, is to say we're all doomed and to give up. In contrast, any story of hope requires much more work than that. And this work, based on compassion and justice and focused on what's good in the moment, might save us. When the doomsayers' predictions overwhelm us, we can still, like Toots, pick up a hammer and start to rebuild. Intentions, actions, and persistence matter.

Our difficulties are predicted to only get harder over the coming decades. As one study indicated, "The world has witnessed a tenfold increase in the number of natural disasters." In 1960, only 39 natural disasters occurred; in 2019, that number jumped to 396.

Through this all, the Udall story becomes more and more relevant. Will we give ourselves to the labor of rescuing and rebuilding? Like most of the town of Udall today, will we be more prepared for the next storm with a shelter close by that can withstand a tornado or hurricane, a fire or flood? And with the people in Kentucky, will we be able to rebuild out of the next floodplain? There will be orphans, along with widows and widowers; will we take them in? We need strong communities, and we will need strong, generous communities to help others when they struggle. The outpouring to both Kansas and Kentucky was impressive, but will that generosity be sustained over the many years of rebuilding in Kentucky and elsewhere? Like in Udall, to lift our spirits and help us remember, we'll need school marching bands and artists of all sorts to help us create and celebrate these new stories. And we will need each other.

Like those Udall survivors, we too are searching through the rubble and wondering how we are going to get through this climate crisis. Toots is yelling at us, *To hell you're leaving*, and he's right, because we have no other planet to call home. Toots is challenging us to tell better stories, to say we'll get through this if we stick together. Any part of the rest of this story, whether it takes place in Kansas, Kentucky, or Cambodia, and any path through the climate crisis will not be easy because the upheaval will be—and already is—great. Yet we have to create narratives that carry us through to thriving, justice-filled, vibrant communities, and these must be stories of love strong enough to say, *To hell you're leaving. I'm rebuilding, and you are* still *going to be my neighbor*.

Left to right: Melanie Turner, Kris Preston, and Sarah Kate Morgan lead carols at the December Gather & Grow event. © Will Anderson

Epilogue

Nickole Brown
Rise

> *Oh, my children, where air we going on this mighty river of earth, a-borning, begetting, and a-dying—the living and the dead riding the waters?*
> —James Still

YOU WATCH AND YOU WATCH AND YOU WATCH.

So you watch and think you know. Yes, you scroll and scroll, just like I once zipped through the channels as a kid—before I had a phone, before there *were* phones like now—and I was watching then as I still do, the false fire flickering across my once-young face, a static blue especially blue these days as I hardly stop watching, can't seem to turn it off.

So someone says *flood*, and you think, *yes, I've seen floods*. Because of course you've seen floods. I know I have—on the news and news feeds, on bigger and bigger screens, some in bars and others in airport gates, one even in a nail salon with a technician kneeling at my feet— flood after flood after flood after—so many now they slop and writhe as one.

But really what you think is a flood isn't *a flood* but *pictures of a flood*, isn't the real deal but its representation, not the signified but its signifier, a complicated experience made gestural, ready to be consumed. Nonetheless, you keep watching, keep clicking through:

TROUBLESOME RISING

Not cars floating but pictures of cars floating. Not waves choking the slender neck of a stop sign but footage of that turbid water, of that red octagon heeded no more. Then, almost always: a woman clutching a drenched and quivering dog. Then, almost always: a man behind her, life-rafting what little he can alongside him in a plastic storage bin.

Underneath it all, predictable captions: *salvage, rescue, damage, toll, rise*. You read the words, understand them, or at least think you do. But then you're distracted by the buzz and ping of all that's incoming. Little red circles and little red bells, one notification and the next, urging you to move on.

Yet that last word you read—that word—*rise*—it rises and rises up the floor within you, brings on a subtle quease, like you're riding an elevator descending too fast.

* * *

I say this because now, I know.

Because now I've lived through. Now, I understand: specifically, a flash flood, in the mountains of Eastern Kentucky, July 27, 2022. Among the forty-five who died: ten in Breathitt County, two from Clay County, and, in the county in which I was located—Knott—twenty-two lost their lives. Among them were four children from one family who clung to a tree until they couldn't no more. Later, I heard that all four were buried in one casket, returned to the earth the same way they died—all together at once.

I was there, in Knott County, teaching for the Appalachian Writers' Workshop at the beloved Hindman Settlement School, having myself a good time eating tomato pie and talking poetry. I was listening to old-time fiddles and sipping bourbon on the porch after supper, and when I strolled back to the dorm, the night was thick with katydids, their sawtooth wings chirring a song that to me *is* July itself.

The flooding began just about then—midweek—a Wednesday—well into the night.

But no, that's not true. It began well before: all that week since we arrived, it had been raining. But not too hard, not really, or so it seemed. Again, I was having myself a good time. Like nearly everyone else, I hardly paid mind to the rain at all, even if it was raining hard.

Forgive me then: I want to show you what a flood is, what it can be. Because though it may not be raining where you are, at least not now, all of us, in one way or another, need to know, need to prepare.

* * *

First, understand the illusion of home, how we believe we're safe because safety is something we need to believe.

Consider the room where I slept: hardly my home, but still, a kind of home—the AC hummed, a clean quilt tucked to my chin, my little dog who'd traveled with me curled and softly snoring at my feet. Across the room, one roommate was sleeping sound, her hands moisturized, her eye mask in place. On her nightstand, the night's glass of water, her bedside reading closed with a bookmark marking where she might pick up tomorrow, confident tomorrow would come.

So when my phone crowed its first weather warning, I wasn't jolted awake, no. I was annoyed. I half-read the word *flashflood* and bemoaned how technology never let me be, thought how silly that something happening far from here should try to rattle a moment peaceful as this. I silenced the alarm, fell back asleep.

About an hour later, our other roommate came in for the night, and though hesitant to wake us, she stood dripping in the doorway and said, *You guys won't believe this, but the creek's rose. My car—it was down in the lower parking lot—and the water's up to my steering wheel.*

Would you believe me if even then I fell back asleep or at least tried to? That all three of us did?

Yes, we might have talked about her poor car or made a joke about the Troublesome Creek living up to its name, how Appalachians hardly mince words if they can help it and no doubt christened that deceptively benign waterway with a suitable adjective. But we weren't worried, no. We agreed: her car was a good distance from us down a steep hill, and besides, it wasn't coming down all that hard. Because, yes, that's the tricky bit: the deeper you're tucked in, the more the illusion persists.

What's funny is even when I *did* get up—after I'd tossed a long while before reluctantly deciding to move my own car to higher ground—I felt silly, feared one of the students might see me traipsing across campus with my pajamas ballooning from my boots, might see me with no bra.

It was only when I got back to the room and realized the water had come ever closer did I begin to sizzle with fright. It was then did I begin to wake. From a distance, the black shape of my roommate's white car—not just flooded but now floating, the hazards flashing as if turned on by a ghost ready to drive across the River Styx.

But even then, I felt ridiculous. I was embarrassed, sure I was being a drama queen. I made fun, said I'd been reading too much Anthropocene poetry, said if there was a national emergency it was just me flash-flooded with hormones, said if they'd humor me and gather their things—*just in case, just in case, I know this is silly, but just in case*—I'd skip out of this dry county tomorrow to a wet one and get them a bottle of their choice. We had a laugh at that—calling the county *dry*. For a minute, our talk mother-eased me, almost had me quit my fretting and go back to bed.

One roommate teased, asked me if I had a catastrophic childhood or something, and I teased back, said, *Well, I told you I grew up in Kentucky, didn't I? Y'all will just have to forgive me for expecting the worst.*

We kept laughing, but all the while, a deeper me began to quake, recalling just how Betty Crocker of a comfort was my childhood home—how sweet and warm the corn bread from the oven, how white and warm the sheets from the dryer—and just how all of that made what happened down in the basement when Mama wasn't looking seem so unreal, so impossible to believe, even now. It stung, that open-handed slap of memories. It made me gather my things even faster.

Before we were done zipping our bags, water snaked under the front door, fast, in long and slender tongues. It was the color of sickness, the weak yellow of bile when a body has vomited everything up but can't seem to quit.

Unable then to open that door and step outside, we lugged our belongings in through the building, through the dank hallway and up the stairs, the water writhing behind, chasing us. The electricity stuttered then stuttered again and quit, so we sloshed through the dark toward the red glow of exit signs.

Once upstairs, alarms beeped on and on, the way they do when someone in a hospital dies, the beeping incessant and unkind until a nurse comes to give the final word and turn off the machines. I flipped every switch I could to make it quit, and when it finally did, I expected silence.

Instead: a groaning, deep and *Titanic*, like a great ship going down.

Because below us, water fast and faster still. Below us, water gurgling up pipes, water bursting glass doors, having its way with the sizable archive stored below, a hundred-plus years of Appalachian literature and photographs destroyed. Below us, our beds lifted and bobbed toward the ceiling. A yellow dress I had hung on the back of the door for the morning and forgot wicked water up to the neck.

The dark made the flood seem loud, and in the dark, it hissed and bubbled. Creaked and blasted. Moaned and sighed. A sound like the dark boiling, like the boiling dark. A sound sounding out a name I will never forget. I tied my dog's leash around my waist, reached to pet her trembling back, felt grateful the door that led outside this floor was still on dry ground.

And that was our first escape from what the Troublesome brought.

* * *

So, then, know this: how water works: how it comes from above as from below.

TROUBLESOME RISING

You see, before last summer, I had what most do of floods: an aerial view, which is why perhaps I thought the velocity was that of rain, falling from above. Yes, aerial—just like on the news—the view from the camera aperture of a drone, seeing as a reporter does, wearing noise-canceling headphones, the helicopter's battering *chop-chop-chop* tossing his hair into a news-at-eleven panic.

But no one truly sees a flood from up there, not really. No, nothing, except for an exhausted cloud or a kiting-high plastic bag. No, no one, except for maybe a bird frantic over her gutter-drowned nest. Except for maybe God, if you believe in *up there*.

Instead: know this: water: how it moves: not only from high to low, but from low to high. How if there's enough of it, it wakes, rises from where it rests, rises from the table from which we take our drink.

And so, as from above as from below: first the floor drains spit up what they were made to choke down, then the toilets and sinks. And the flooring: first just slightly cool under our feet, a dampness easily dismissed. Then, a wet stain phantoms up. Finally, all of it—the rugs and carpets, the tiles and floorboards—they, too, lift and float.

I say this because the rain that week was comforting, seemingly benign. At times, it was a snuggle-down, book-reading mist. Or an easy, window-glazing kind of drumming. I mean, consider this: my roommate—even after her car floated away—she curled into the sill, said, *Rain like this always soothed me. Ever since I was a child, it just makes me feel calm.* And the other roommate, she later wrote to say the rain that evening didn't look menacing to her, just odd. *So I took a video of it*, she said, adding, *But who takes videos of rain?* Well, I'm sure lots of people do, but it never occurred to me before. A form of praise, I think.

But when it floods, the word *water* tendrils, morphs into *waters*, that single letter—that *s*—makes water plural, that singular noun makes not one unified water but water from all directions, across from rivers and up from wells and down from clouds—*water* made *waters*—running down mountains and melting off ice caps, making creeks of ditches, rivers of creeks, lifting the good earth beneath our feet and making it slide. *Waters*, rising and rising, without cease, like a bathtub running and forgotten, like a bath running for someone slumped within it, their wrists slit.

Waters sounding somehow more poetic, more holy. *Waters* like a body letting go of its water, muscle made mud, eyes made mud, skin and hair and organs—all mud, mud, mud—a body returning to the thing from which it was made.

Waters, sounding downright biblical, ready for a baptism maybe and given power, which is what water is.

Long after, I'll take a sip of coffee, think what if this is not coffee but water that remembers it has passed through something, that it's been changed, stained by me and a scoop of grounds? What if this were not *coffee* but *water*, which truly, it is? What if it was not *water* in my mug but *waters*, what does that mean? I take another sip, think this is the kind of shit that keeps me from getting anything done, the quivering I can't seem to quell, even months after that flood. I shake it off, answer another email. But still, a part of me never stopped wading through that ceaseless night.

* * *

Because a flood, it will lodge in your body; it will change those who wade through, will call to task what your eyes think they know, will have your ears and nose summon the jagged rest.

Thus, the smell: High electric, an ion-charged whine, a seething scream of petrol and piss, a way-up-in-the-head stench, migraine-tinged, lit by the kind of fluorescents in places where you'll see things you don't want to see. Worse, the smell is not just sewage and gasoline but is made discordant by a smell you love: petrichor, that earthy goodness of summer showers, the microbes long kept in the soil rising up again.

What I mean is the smell is part earth, part what we have done to the earth.

TROUBLESOME RISING

What I mean is the smell is part miracle that made and sustains us, part what we have done to destroy that sacred covenant, what we have likely done to cause this mess. Once you breathe it, it will reside within, your lungs watermarked like the walls of a once-flooded room now dry but under that fresh sheetrock a discolored line that never forgets.

Worse, it's a smell made visible by what's caught on the fence line: plastic bottles of detergent and plastic bottles of antifreeze and plastic bottles of hand sanitizer and plastic bottles of drinking water looking ironic as hell floating in all that water. And there, in the limb of a tree, even a plastic bottle for a baby, still somehow half full of milk.

Later, it's a smell doubled by things you can't first identify or even fathom that come to knock down the fence and the power line and whatever else is in the way—a floating parade of feed troths and dumpsters and pickup trucks, porch furniture and couches. Later, toolsheds and trailers, the wet rag of a dead raccoon, the limp hose of a dead snake. Then, a thing that makes you laugh a little but in shame: an above-ground swimming pool, its beach ball pinwheeling green and pink and yellow in the fetid rush.

Further down the road, there's a house lifting from its foundation that crashes into another. They say those two homes then lift together and rush into another—three homes made gruesome dominoes that freight-train into an eighty-something woman who lived her natural born in these mountains until that night.

* * *

And know the flood won't just lodge in your body. No, your mind, too: it will split.

Because not far down that first dark hallway, I could feel it happen: while a part of me was ever present, another was far into the future, remembering already what was happening now.

Yes, you will split:

A part of you will be pure amygdalae, keenly aware, all your senses tinged with adrenaline, but at the same time, another you will cling all at once to the past and the present and the future, desperate to make sense of what's happening before it's even over.

Put another way, you'll scramble to figure how, if you live, you're going to tell this story.

No, that's not quite right: you'll feel if you *don't* figure a way to tell this story, you *won't* live.

No, let me try this again: you'll tell the story to yourself in past tense as it happens, minute by minute, each moment already translated into the future, already a story you've known your whole life.

Consider the walk to the next shelter we could reach at the top of the campus's main hill—an old house made into a dorm. In my hand, a book light no bigger than a fairy's lantern to illuminate each step, my boots disappeared in that khaki-colored rush, my feet alert to every tug and vibration, feral with the pock and tap of every small rock and stick charging past.

At the same time, before I got to the door, I was already at the door, had already locked and loaded my line, saying, *Have room for a climate refugee?*

It was meant to be a wry joke, the kind of line I'd want the actor playing my part in the movie to say, a commentary on this terror weather told in a lighthearted way. But then I thought, *No, how arrogant, how Little Miss Privilege of me. . . . A frightening evacuation from one building to the next hardly a refugee makes. Besides, you don't even live here; your home is a hundred miles east of here, safe and sound.*

I rewound the film, edited out that scene, shot the footage again. I said that line, then unsaid it, all while walking through the door I had not yet reached.

TROUBLESOME RISING

So I opened the door and open the door and will open the door, all at once; I coaxed the verb *open* to open to all tenses together and do just that to the door.

When I do, the door opens to a silent and sleepy-time dark—there is no one to even receive anything I could say. They are all still in bed; none of them even know what has happened. None of them know what's happening now. They don't even know half their vehicles have washed away.

Hello??!!?, I said, drawing out that knock-knock-anybody-home vowel of *o*.

Hello??!!?, I say, making a fist and using it on one door then the next.

Hello??!!?, I will say, again and again, into that dark forever, waking each person, telling them to rise.

Eventually, all of us gather together in the pitch black with no idea what to do as great waters form around us and roar past. We tremble to think the steep incline above us will slide, knowing there is nowhere else to go.

I feel guilty—there is nothing any of us can do. I should have let them sleep. It's only when one dreams that all tenses should operate like that at once.

* * *

Here then is another thing to know: when it happens, when a flood comes, you won't act the way you think. You might not even notice the flood has washed the you that was you away.

See then the motley huddled dozens of us, together in the dark:

One crumples and cries, murmurs something about a fire. Another stumbles backward, sloshing liquor to the floor. Another blasts the room with a machine-gun fire of *fuckfuckfuckfuck*, then chucks in a grenade of *goddamnit, Appalachia, goddamn!* Another runs outside to try to save her motorcycle and returns defeated, saying she had just paid for it in full. One walks around with a Sharpie, telling us to write our names on our arms just in case our bodies need to be identified. Someone else paces, another falls asleep snoring upright in a chair, someone else says, *Well, hell's bells, I'm going to have a smoke then, because what difference does it make now if I quit?* Another distributes blankets. Another gathers the blankets and puts them up. Another passes the blankets back out again. One stares as a hawk about to strike prey; another stares as a horse dozing in the sun. Yet another tells us of a flood she escaped as a kid, how her mama in a frenzy to evacuate gathered only three things: money from the dryer, a bunch of bananas, and her girdle. I have a good laugh at that, deeply thankful for the kind of humor Kentucky brings.

And me? I churn petty anxieties, chide myself for what I left behind in the room: my dog's treats, my dental floss, that yellow dress. I open my bag, check to see if I remembered to get my hoodie, my notebook, my computer, my wallet, my phone. I let my dog off leash so she can perhaps comfort people, then I panic, worried I won't be able to find her if the hill above gives way. I clip her in again, double-knot her leash to my waist. I check for my hoodie once more. I check for my hoodie again, my notebook again, my wallet, my phone. I check for my jewelry and then put my jewelry on, careful to secure the heavy bracelet my mama bought me when I was but twenty-one, finally facing those dank early memories I had tried so hard to erase. Instead of taking my own life that day, I went shopping with her instead, and for over two decades, I've worn that bracelet, wielding it like some Wonder Woman kind of cuff. I put it on my right wrist, figure it will either be heavy enough to keep me from floating away or identify my body when they find it.

More and more people from the lower buildings arrive, filling the crowded dorm so much that we spill out onto the covered porch. A cat yowls from her crate. A little wet dog yips and shivers in the bathroom, and when my dog sees him, it's as if she couldn't care less about meeting other dogs anymore and pays him no mind. Again, I tighten her leash around my waist, figure if we're carried off at least we'll be together. Figure if I need a length of rope to tether the two of us to a tree, I can use that.

TROUBLESOME RISING

Not much later, a friend leans in, says, *You know, the land above us is forested; the ground above us should hold.*

It's then I stop my spiraling, take the needle off the record going round and round in my head.

Much of the land here has been logged completely, he adds. *But the trees above us should hold. But we won't know until morning.*

I can't remember saying anything in response, but to myself, I thank the trees for holding and thank whoever decided to leave those trees standing, especially knowing just how hard money is to come by in these mountains and for how much money such trees could be sold. Then, from deep down comes a scrap of lines from a half-remembered poem:

Earth loved more than any earth, stand firm, hold fast; / Trees burdened with leaf and bird, root deep, grow tall. Mister Still—as folks still call him around here—he wrote that.

Because yes, how much land here has been logged? I've never lived in Eastern Kentucky, but still: all those stories about my great-grandmother feeding lumberjacks out this way, how she'd fry up twenty-one chickens at a time for those men, just for lunch, how she lived in poverty that ground the word *grinding* to dust. And just how many trees did those hungry men cut? How much coal did they crawl into the earth to pick and haul? How many trains did it take to carry all of that far away from here, to build and heat cities full of people who would look down on this place?

Yes, how much money those trees could earn, how much even one tree might cost. How many lives huddled below those trees that night.

Even now, I repeat Mister Still's words, like a mantra, again and again and again: *Earth loved more than any earth, stand firm, hold fast; Trees burdened with leaf and bird, root deep, grow tall.*

Again, say it with me: *Earth loved more than any earth, stand firm, hold fast; Trees burdened with leaf and bird, root deep, grow tall. . . . Earth loved more than any earth, stand firm, hold fast; Trees burdened with leaf and bird, root deep, grow tall.*

More than any earth. Hold fast. The trees, burdened. The trees, left to grow, to root deep. Earth, firm, deep, more than any earth. Bird, loved. Trees, loved. Earth, loved, hold, hold, hold. . . . Toss the words from those lines around any way you want, and you will still find a prayer there worth praying.

* * *

Know too there will be an *after* that comes after. There will be more to follow; the story won't just end after you're too tired to tell the story to yourself. No, it will go on and on like a story you don't want to watch anymore but can't turn off.

Which means even though this night will end, the story will not. Which means instead of morning like you've always known, the first light will reveal not the return of day but a day turned in on itself, a day torn from the one that came before.

Which means at first light will be four ducks in front of the dorm, and though amused by their waddling, you realize they're clearly domestics and are searching back and forth and back and forth to take shelter inside a barn that's far from here and likely no more. Eventually, they'll stop their panic and will curl exhausted next to each other, their fresh-snow necks tucked into their fresh-snow backs, an unreal spot of brilliance on that mud-slaked hill.

Which means too that at first light you'll walk to what's left of downtown, amazed the little convenience store—the aptly named Mi-Dee Mart—is still there. You'll stand at the edge of a furious brown river that yesterday was Main Street. Next to you, a man smokes the way working men you've known all your life do—arm down, hiding their cigarette at their side, using their palm to protect the glowing cherry from the wind, smoke rising from their open fist. *Never in all my forty-five years have I seen such a sight*, he says, not to you exactly but meant for you to hear. Next to him, another man, smoking much the same country way, replies, *No, never, not in all my twenty-five years*. And yet another, exhaling a long time as if he's pushing smoke out of his lungs though he's not smoking a thing, says, *No, never. Not in all my seventy-three.*

When you return to the dorm, it's decided that those who still have cars to speak of should try to get out of there while they can. Besides, more flooding is predicted to come, and the Settlement School needs all the housing they have left to help those who live there. You'll never forgive yourself for this, but you do as you're told: You leave quickly, without so much as a good-bye. You flee without staying to see the waters recede; you take off without helping to haul so much as one box of precious archives out of the mud. You don't even stay long enough to see if the owner of the four little white ducks will come and take them home.

And here is where you'll have to forgive me for speaking in third person, because this was *me*, not *her*—I know—but the evacuation is too much to bear with blatant capital *i*'s as if *I* were there. Because, yes, telling this next part in first person would be a lie. So I'll tell you the rest in third because the me that is me was, by then, long gone, because driving away that day was a stranger—a stranger not *to* me but a stranger *of* me—an artificial intelligence that felt nothing as she listened to what the me that was left of me tried to say.

Because all that day, she tries and tries to get back home, but there's more mud than pavement, more water than road. She hydroplanes through one intersection then the next, and when the road is impassable, she turns the car around, hydroplanes through the same intersections all over again. She rides the rumble strip on the side of the highway, and later, she turns around yet again and rides that same rumble strip once more, this time on the wrong side of the road.

The gas pumps will be closed; the gauge will nearly tick down below *E*. There's no place to stop for food. She'll come to one dead end and then the next and the next, not sure if the water gets deep enough if it's better to shut the engine off or gun it or what. She'll pass one destroyed home after the next; she'll dodge dislodged trees and stones; she'll dodge mattresses and clothes and other cars not as lucky as hers.

Right outside the town of Hazard, she'll pass a brown dog twisted into a pile of meat, and unbelievably, a swallowtail will be drinking from the side of his wet neck. She'll see that dog again and again; I even see him sometimes today—the twist of his open muzzle, the heap of red, that flitting yellow pumping its bright wings.

Unable to find a way east over the mountains to get back home, she'll finally turn around, head west instead, away from where she doesn't belong now anyhow, away from home.

Exhausted, she'll stop in central Kentucky, and there she'll find an empty room empty of everyone except her dog and herself. The room will be basic and cheap; it will be blessedly clean, blessedly dry. When she sees the white of sheets crisp on the bed, she won't weep though she wishes she could. She won't bathe. No, she won't even take off her rain boots. She'll slump her filthy body into that clean bed and won't even remember falling asleep.

The next afternoon, she'll wake. In the mirror, she and her dog both will be slaked with dried mud that comes off in flakes and drifts; there will be a fine dusting of red-brown on everything. Her pajamas will still be tucked into her boots, and she still won't be wearing a bra. She'll wonder at what point she went from being ashamed the night before last to not noticing how she was dressed at all.

On the counter is the kind of coffee machine with a filter tray made of plastic so disposable it melts and falls off when the hot water runs through. Little plastic bags of napkins and little plastic bags of plastic stir sticks. A tiny bag of powder to turn her coffee pale and pretend it's cream. A tiny bag of sweetener with a cancer warning on the side. A small sign that states the grounds are 100 percent rainforest select. She makes a cup but can't drink it; the weak brew's the color of that water she just left behind.

She'll pour out the coffee. She'll shower, crawl back into bed. It's there she'll stay for two days.

* * *

TROUBLESOME RISING

Finally, know too how once you turn on the television to watch the news, how what you've just survived will somehow seem less real, even to you, how it will slide back into the familiar trope you've seen so many times before. Know how much work it is to hold on to what the body knows, how the media will try to tell you how to feel in between commercials about pills for depression and pills for blood pressure, commercials for pizza and burgers and under-twenty-minutes-or-less-easy-breezy meals.

Worse, as the world is made as it is now, it also demands that you live two lives—one in person and another online—so you pick up your phone and scroll through to find another foul current, this time made of words:

These people got what they voted for, one says. And yet another: *We should just let them swim.* And another: *What are those houses doing there along the river in the first place?* And another: *Maybe it's God's punishment for being a bastion of ignorance and regression.* And another: *I've never seen so many banjoes floating down the river.* And another, and another, and another, so much so I wade through another long darkness in a different kind of flood.

I also watch clips of people from the drowned counties, hear in their talk the accent that once thickened my own tongue, the kind of talk that somehow others think it okay to imitate with a *Beverly Hillbillies'* yee-haw, that has them hum the notes from *Deliverance* in case I didn't get the joke. Yes, even my most educated and liberal friends make redneck jokes to my face and don't think a thing of it.

We didn't have no time to get nothing out, one woman says in one clip, her heartbreak of a double negative shaking its head *no* and *no* and *no*.

We're not gonna let up till everyone is accounted for, says another, and I think of how I once cut my teeth on the word *kin*, far before anyone ever uttered the word in any academic circles, how I was brought up to think blood thicker than any water, even floodwater thick with blood itself.

Later, a journalist will call to get a statement from me on the phone and will say what he can't say in print: never in all his years of reporting flooding in Eastern Kentucky has he seen anything like this. Regardless, if he says the words *climate change* in his article, what kinds of hate mail he might get.

Somehow, it makes me recall an old joke I've heard more times than I can count, one about a hillbilly stuck on his rooftop in a flood, crying out to Jesus to save him. Soon, a man in a rowboat comes by, says for him to jump on in, but the man refuses, says he was praying so hard that Christ Himself would come down from the cross and save him. Then a motorboat comes by, offers the same, but the man says no, that he's righteous with a snakebite kind of faith, says he sure doesn't need the likes of any help from mankind. Finally, a helicopter buzzes down, and the pilot shouts for him to grab a rope and be lifted to safety. Again, the man refuses, raises his arms, speaks in tongues instead.

Perhaps you've heard that one already? If not, maybe you've heard a simplified version of the same? It's well known enough, and more than one person told that joke while we waited for the sun to rise the night of the flood. Either way, I bet you can guess the punch line: The man drowns and goes to heaven. Once there, he asks God why his prayers weren't answered, why he just let him die like that. To this, God replies, *Well, I sent you a rowboat and a motorboat and a helicopter. . . . What more did you expect?*

What I can't help wonder, though, is this: What if there's much missing from the joke? What if that first would-be rescuer, the one in the rowboat, made fun of the way the man talked, figured that helping him was a charitable thing in a better-than-thou sort of way? And what if the owner of the motorboat was trying to coax the man off his roof in exchange for the rights to his land, figuring there was a good seam of coal gleaming under all that brown water, that those old-growth trees would be that much easier to cut after they died back from the flood? And what if the helicopter was just another government handout, there to lift the man up for a spot on the news before ditching him in some arena crowded with evacuees where he'd have to work that much harder to even think of finding his way home?

Worse, what if none of that was true—what if each person who offered help was as good intentioned and kind as they come—but the man, having sprung from generation upon generation of exploitation and ridicule, simply couldn't trust the likes of any of them and instead put all his faith in God?

This, of course, is the kind of pissed-off thinking that gets me in trouble, that keeps me from taking pleasure in a simple joke.

* * *

So now I know. And maybe now you know or at least know more than you did.

So come with me then, back to the moment before it began, before I didn't know, back to that *before* to which I return, again and again: to that last night I had in Hindman, in the eastern part of the state, at that beloved settlement school there. Again, you know the story by now: It's Wednesday night. It's raining. Of course it is, but we barely notice.

Yes, it's raining, but it's not a thing that worries us, at least not yet. Instead, we listen to katydids thicken and texture the air—that sound that *is* Kentucky in July itself.

If you would then, slow with me a spell. Let the song enter us; allow that sugared darkness into you, that sweet dark that frightens and soothes at once. Because in that insect song is something that says despite all the words we read and words we spoke and words we heard today, we're still animals just as my dog is an animal and the katydids are animals though my dog is sitting at my feet and looking up to ask us why her walk has stopped.

Please: Stay with me, just a beat longer, not so much *listening* to the katydids as we *drink them*, with our ears, and let me tell you an Appalachian tale I half-remember—about how theirs is a call-and-response song of *Katy did, Katy didn't . . . Katy did, Katy didn't . . .* something about a girl named Katy who killed somebody—maybe her sister, maybe a man she loved, maybe both—but whoever it was, the katydids way up in the branches saw everything like they always do and now have something to say.

We maybe laugh then about stories from these mountains and how they almost always involve some hideous crime of passion—*nothing better than a murder ballad sung by bugs*, of all things—and we'll listen to the insect jury debating on and on in the trees, back and forth and back and forth.

Nonetheless, let me tell you: whatever Katy did, she must have done eons ago, because the insect song is so old within me that I remember the song without having any memory of it, as if my first memories of hearing them didn't just come after I was born but somehow were born *with* me.

And you? I have to ask: Do you ever feel the same? Do you ever remember something from so deep down that it seems like something you were born into, that you were born knowing, before you even knew what knowing was?

I know, I know. Maybe I've had a touch of bourbon and should quit my rambling and tuck in for the night. Besides, it's raining harder now. It's time for a little shut-eye.

But you know the story now, don't you? You know this night won't end, not now, not for a long while to come. Because though we don't notice, at least not yet, the flooding has begun.

But no—that's not true, not exactly—the flood—it began far before us, didn't it? Maybe even before we were born. It was born in us and with us, going generations back when prospectors first opened up a vein of coal in these mountains, back when the first truck sputtered up one of these winding roads carrying a load heavy with timber.

And the katydids? Maybe they couldn't care about poor Katy and her jealous heart. Maybe, just maybe, instead of talking about her, they're debating on and on about us, trying to decide if what's about to happen is a crime and if it is, if we're to blame.

Forgive me. Maybe you could care less about insects. Maybe you think that my saying all this nonsense about katydids is just another human projecting a human story on a creature doing their own thing for their own reasons. And you're right.

Nevertheless, do you hear them? What happens when you cock your head to the side and listen to what this earth—and its waters—has to say? Or are you asleep by now like I was, tucked in my bed?

If so, just know it won't be long before you might hear a pounding on your door calling out, *Hello*??!!?, just as I did, drawing out that knock-knock vowel of *o*.

No, what I mean to say is this: Are you listening? The waters are coming. Even if it's not raining where you are, even if you've been in a drought for years and years, even if you shower with a bucket at your feet to catch the soapy runoff and lug it outside just to keep a few struggling plants alive.

Forgive me if I say, *Hello*??!!?, knocking on one room, then the next. There may be nothing any of us can do, but please, turn off your screens, open your door, ask the land if it will hold when the waters come. If not, plant tree after tree after tree; in years to come, they may just be all that's between a house of terrified people and the weather-terrorized land about to slide.

Hello??!!?, I will say, again and again, for as long as I'm able, begging every person I can, telling them to rise. For mercy. For the earth loved more than any other earth. Rise, please, rise. You must wake now. We must rise.

A Final Note

BECAUSE THE WRITING FAMILY OF HINDMAN SETTLEMENT SCHOOL and the community surrounding the campus who experienced the flood are both so broad, only the smallest gathering of poems, stories, essays, photographs, and experiences could be included in this book. For that reason, the settlement school has established a companion website for the anthology: TroublesomeRising.org. With this website, we want to include more stories, voices, and experiences connected to this historic flood. We want the experiences of what happened during those early days of rescue and recovery and also later. What will we have to say about it five years from now? Ten? What will our kids say about it once they are grown? What will we have to say as we face discoveries in research and changes in climate, politics, and economics? We want this website to not only expand the documentation of the flood and act as a more inclusive archive than the anthology can on its own but also act as a portal through which the broader world can learn about what happened here—and learn about it from our own words.

Acknowledgments

MANY THANKS TO THE EDITORS OF THE PUBLICATIONS IN WHICH some of these works first appeared, sometimes in slightly different forms, and to those who have graciously agreed to share their words here for the first time.

Berry

"Making It Home" was originally published in Wendell Berry, *Fidelity: Five Stories* (Berkeley, CA: Counterpoint, 2018). Reprinted by permission of the author.

Brown

James Still, excerpt from *River of Earth* (Lexington: University Press of Kentucky, 1978), 76. Used by permission of the University Press of Kentucky and the estate.

"Rise" was published in *Elementals: Water*, vol. 3, ed. Ingrid Stefanovic (Libertyville, IL: Center for Humans and Nature Press, 2024). Used by permission of the author and the Center for Humans and Nature Press.

Hague

Quote and verses by SAWC volunteers reprinted by permission of the authors.

Acknowledgments

Hansel

"Aerial View of Catastrophic Flooding in Eastern Kentucky" was originally published by the *New Verse News* and is made up of direct quotes from online posts about the flooding. Pauletta Hansel, "Aerial View of Catastrophic Flooding in Eastern Kentucky," *New Verse News*, August 2, 2022, https://newversenews.blogspot.com/2022/08/aerial-view-of-catastrophic-flooding-in.html. Reprinted by permission of the *New Verse News*.

"No Friends of Coal" was originally published by *The New Verse News*. Pauletta Hansel, "No Friends of Coal," *New Verse News*, September 6, 2022, https://newversenews.blogspot.com/2022/09/no-friends-of-coal.html. Reprinted by permission of the *New Verse News*.

House

"Pulled from the Flood" was originally published by *Garden & Gun*. Silas House, "Pulled from the Flood," *Garden & Gun*, August 3, 2022, https://gardenandgun.com/feature/pulled-from-the-flood/. Reprinted by permission of *Garden & Gun*.

McCarroll

bell hooks, excerpt from *Appalachian Elegy: Poetry and Place*, p. 67. Copyright © 2012 Gloria Jean Watkins. Used by permission of the University Press of Kentucky and the estate.

Lucille Clifton, excerpt from "cutting greens" from *The Collected Poems of Lucille Clifton*. Copyright © 1974, 1987 by Lucille Clifton. Reprinted with the permission of the Permissions Company LLC on behalf of BOA Editions Ltd., boaeditions.org.

Minick

"After the Flood, After the Tornado, and Before the Next" adapted from *Without Warning: The Tornado of Udall, Kansas* by Jim Minick by permission of the University of Nebraska Press. Copyright © 2023 by Jim Minick. Used by permission of the University of Nebraska Press.

Acknowledgments

Reid

An earlier version of "What We Saved" was published in Erin Reid, "'What Could I Have Saved?': Eastern KY Floods Took Our Present, but Also Our Past," *100 Days in Appalachia*, December 19, 2022, https://www.100daysinappalachia.com/2022/08/what-could-i-have-saved-eastern-ky-floods-took-our-present-but-also-our-past/. Used by permission of *100 Days in Appalachia*.

Sheffel

An earlier version of "What Water Can't Erase" was published in Mandi Fugate Sheffel, "We Will Rebuild in EKy. Then We Must Ask Why 100-Year Floods Are Happening So Often," *Lexington Herald-Leader*, July 31, 2022, https://www.kentucky.com/opinion/op-ed/article264016006.html. Used by permission of *Lexington Herald-Leader*.

Sickels

An earlier version of "Troublesome Rising" was published in Carter Sickels, "The Historic Kentucky Floods Were a Waking Nightmare—and They're Only the Beginning," *Outside Online*, October 14, 2022, https://www.outsideonline.com/outdoor-adventure/environment/eastern-kentucky-flooding/. Used by permission of *Outside Online*.

Appendixes

Appendix 1

The map on the following page shows the locations of key buildings on the Hindman campus. Map by Tyler Barrett.

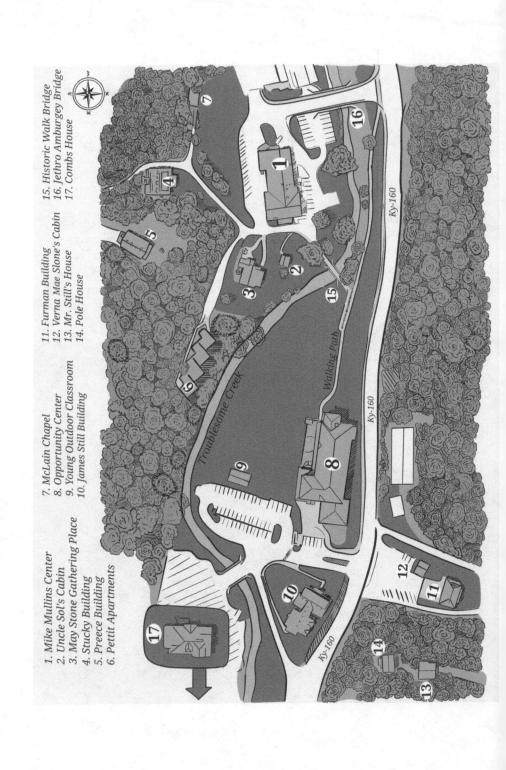

Appendix 2

45th Appalachian Writers' Workshop Schedule
July 24–29, 2022

Sunday, July 24

3:00 p.m.	Registration	Gathering Place
6:00 p.m.	Dinner	Dining Hall
7:30 p.m.	Orientation and Introductions	Great Hall
8:45 p.m.	Screening under the Stars: *The Evening Hour (Carter Sickels)*	Hindman Amphitheater

Daily Program

Monday–Friday

*Monday–Thursday Only

8:00 a.m.	Breakfast	Dining Hall
9:00–10:30 a.m.	Concurrent Workshop Sessions	
	Poetry—*Nickole Brown*	James Still—Upstairs

Appendix 2

	Creative Nonfiction—*Meredith McCarroll*	James Still—Downstairs
	Novel—*Caleb Johnson*	Gathering Place
10:45 a.m.–12:15 p.m.	Concurrent Workshop Sessions	
	Poetry—*Lyrae Van Clief-Stefanon*	James Still—Upstairs
	Short Story—*Jayne Moore Waldrop*	James Still—Downstairs
	Novel—*Carter Sickels*	Gathering Place
12:30 p.m.	Lunch	Dining Hall

1:45–3:00 p.m. Afternoon Sessions*

Monday
Generative Session: Picture Books—*George Ella Lyon*
Songwriting—*Britton Patrick Morgan*

Tuesday
Nature: Medicinal Muse—*Bernard Clay*

Untold Stories: Hindman, Historical Fiction & Hope—*Patricia Hudson*

Wednesday
Publishing Panel—*Ashley Runyon, Patrick O'Dowd & Abby Freeland*

Thursday
Generative Sessions: Poetry—*George Ella Lyon*
Songwriting—*Britton Patrick Morgan*

3:15–5:15 p.m.	Participant Readings* (M/Tu/W)	Great Hall
4:00–5:15 p.m.	Creating Characters: (Th) *Silas House in conversation with Little Bubby Child*	Great Hall

5:30 p.m.	Dinner*	Great Hall
7:00 p.m.	Evening Program*	Great Hall
8:30 p.m.	Techno Contra Dance* (M)	Great Hall
	Troublesome Trivia* (W)	Great Hall
	Book Signing & Reception* (Th)	Great Hall

Evening Programs/Specials

Monday, July 25 — Nickole Brown & Jayne Moore Waldrop & Caleb Johnson

Tuesday, July 26 — Britton Patrick Morgan & George Ella Lyon

Wednesday, July 27 — Meredith McCarroll & Carter Sickels & Lyrae Van Clief-Stefanon

Thursday, July 28 — Jim Wayne Miller/James Still Keynote Address with Beth Macy Book Signing and Reception to Follow

Informal gatherings are encouraged each evening after the final scheduled activity of the day. These will take place at various locations all over campus and are up to individuals to organize. Our musician-in-residence will gather people at the Gathering Place each evening. The Settlement School desires people to get to know one another better during these times, but the gatherings are completely optional and will not be organized by Settlement School staff. They are not intended to be exclusive gatherings but informal gatherings for the purpose of inclusive fellowship.

Appendix 3

The figure on the following page shows the evolution of the Hindman Settlement School over time.

HINDMAN SETTLEMENT SCHOOL
historical timeline

1899 — Knott County local, "Uncle" Solomon Everidge asks May Stone and Katherine Pettit to found a school "for his grands and greats."

1902 — Hindman Settlement School is founded.

1910 — Hindman Settlement School becomes Knott County's official high school.

1921 — Hindman Settlement School becomes an "approved school" of the National Society Daughters of the American Revolution.

1942 — The Settlement begins extension education program in remote county schools.

1977 — A folk dance and writers' workshop is held, which marks the beginning of the Appalachian Writers' Workshop and Appalachian Family Folk Week.

1979 — Hindman Settlement School begins hosting tutoring sessions for children with dyslexia, which launches the Dyslexia Education Program.

2013 — The Settlement launches Knott County Grow Appalachia, laying the foundation for its work in agriculture and foodways.

2018 — The Mike Mullins Cultural Heritage Center and May Stone Gathering Place open on campus.

2020 — An expansion of the Dyslexia Education Program into 16 additional schools across the region begins through the Reading Corps project.

2022 — Inaugural residential writing program for high schoolers, Ironwood, is launched.

Contributors

G. Akers is an English education student at Morehead State University whose work has been published in *Still: The Journal* and Morehead State University's literary journal, *Inscape*. She hopes to continue learning and writing and plans to dive deeper into the rich heritage of Appalachia and the people around her.

Will Anderson has been a staff member at Hindman Settlement School since June 2021. He is a self-taught photographer whose work has appeared in numerous publications and news media, with a focus primarily on life and Native culture in Alaska.

Darnell Arnoult is the author of the novel *Sufficient Grace* and two poetry collections, *What Travels with Us* and *Galaxie Wagon*. Her shorter works have appeared in a variety of journals and anthologies. She is the recipient of the SIBA Poetry Book of the Year Prize, the Weatherford Award, the Chaffin Award, the Hobson Award in Arts and Letters, and a Pushcart Prize nomination. She lives with her family in Mebane, North Carolina. For more information, visit darnellarnoult.net.

Neema Avashia was born and raised in Cross Lanes, West Virginia. She is a longtime educator in the Boston Public Schools and the author of *Another Appalachia: Coming Up Queer and Indian in a Mountain Place*. She lives in Boston with her partner, Laura, and her daughter, Kahani.

Tyler Barrett was born and raised in Albuquerque, New Mexico, and has lived in Hindman for three years. He has been an artist his entire life, working professionally in various mediums since 2015. After volunteering during flood recovery efforts, he started work at HSS as the Community Engagement Coordinator and has since built a running series of community outreach events called Gather & Grow, designed to reunite and reconnect folks affected by the flood. His art/photography can be viewed on Instagram @ch3x_mix.

Contributors

Wendell Berry is an American novelist, poet, environmental activist, cultural critic, and farmer. A prolific author, he has written many novels, short stories, poems, and essays. He is an elected member of the Fellowship of Southern Writers, a recipient of the National Humanities Medal, and the Jefferson Lecturer for 2012. He is also a 2013 Fellow of the American Academy of Arts and Sciences. Berry was named the recipient of the 2013 Richard C. Holbrooke Distinguished Achievement Award. On January 28, 2015, he became the first living writer to be ushered into the Kentucky Writers Hall of Fame.

Nickole Brown is the author of *Sister*, a novel-in-poems, and *Fanny Says*, a biography-in-poems about her grandmother that won the Weatherford Award for Appalachian poetry. She lives in Asheville, North Carolina, where she volunteers at several animal sanctuaries. *To Those Who Were Our First Gods*, a chapbook of poems about these animals, won the 2018 Rattle Prize, and her essay-in-poems, *The Donkey Elegies*, was published by Sibling Rivalry Press in 2020.

Wes Browne is an author, attorney, and restaurant owner. His debut novel, *Hillbilly Hustle*, released in 2020, was named one of Merriam-Webster's seventeen recommended pandemic reads. He lives with his wife and two sons in Madison County, Kentucky.

Annette Saunooke Clapsaddle is a citizen of the Eastern Band of Cherokee Indians. Her debut novel, *Even as We Breathe*, is the first novel published by a citizen of her tribe. It was a finalist for the Weatherford Award, was named one of NPR's Best Books of 2020, and received the Thomas Wolfe Memorial Literary Award. Clapsaddle's work appears in *Yes!* magazine, *Lit Hub*, *Our State Magazine*, and the *Atlantic*.

Kentucky native **Bernard Clay** grew up in Louisville. Bernard earned an MFA in Creative Writing from the University of Kentucky Creative Writing Program and is a member of the Affrilachian Poets Collective. His work has appeared in *Appalachian Heritage*, *Limestone*, *Blackbone*, and *Once a City Said*. He now lives in eastern Kentucky on a farm.

Monic Ductan lives in Tennessee and teaches writing at Tennessee Tech University. She is the author of the story collection *Daughters of Muscadine*. Her writing has appeared in numerous journals, including *Kweli*, *Shenandoah*, *Oxford American*, and *Appalachian Review*.

Nikki Giovanni, as a little girl, sat in the window of the bedroom she shared with her older sister and read by flashlight. She looked at the stars when the battery gave way and snuggled under her grandmother's quilts to listen all night to jazz on the radio. She first fell in love with words, then they somehow seemed to fall in love with her. She got to learn history, meet people, and travel everywhere; and since this is a good fairy tale, she lives happily ever after, becoming an award-winning poet, author, and civil rights activist. There may be other things along the way, but the words and the stars and the music are all that matter.

Robert Gipe is the author of three novels and founding producer of the Higher Ground community performance series. He lives in Harlan County, Kentucky, and grew up in Kingsport, Tennessee. He is the author of the illustrated novels *Trampoline*, *Weedeater*, and *Pop*.

Elizabeth Lane Glass has a PhD in Humanities from the University of Louisville. She has received an Emerging Artist Award in Nonfiction from the Kentucky Arts Council and a grant from the Kentucky Foundation for Women. Her writing has appeared or is forthcoming in publications such as *Redivider*, *River Teeth*'s Beautiful Things series, *Redefining Disability*, and *The Manifest Station*. She lives in Louisville, Kentucky.

Jesse Graves is the author of four poetry collections, including *Merciful Days*, and a collection of essays, *Said-Songs: Essays on Poetry and Place*. His work received the James Still Award for Writing about the Appalachian South from the Fellowship of Southern Writers and two Weatherford Awards from Berea College. He teaches at East Tennessee State University, where he is poet-in-residence and professor of English.

Kari Gunter-Seymour is the Poet Laureate of Ohio, an Academy of American Poets Laureate Fellow, a Pillars of Prosperity Fellow for the Foundation for Appalachian Ohio, the founder/executive director of the Women of Appalachia Project, and editor of its anthology series *Women Speak*. Her current collections include *Alone in the House of My Heart* (2022) and *A Place So Deep Inside America It Can't Be Seen* (2020), winner of the 2020 Ohio Poet of the Year Award.

Richard Hague has taught and led workshops for the Appalachian Writers' Workshop, the Tennessee Mountain Writers Conference, the Writers Conference of Northern Appalachia, and the Highlander Summer Conference at Radford University, among others. The 2004 winner of the Appalachian Poetry Book of the Year (*Alive in Hard Country*, Bottom Dog Press) and the 2013 Weatherford Award in poetry (*During the Recent Extinctions*, Dos Madres Press), he currently writes and leads workshops for the Originary Arts Initiative.

Leah Hampton is the author of *F*ckface: And Other Stories* (Henry Holt). Her work has appeared in *storySouth*, *Electric Literature*, *McSweeney's Quarterly*

Contributors

Concern, Appalachian Review, Georgia Review, and elsewhere. She is assistant professor of creative writing at the University of Idaho and divides her time between the Pacific Northwest and the Blue Ridge Mountains.

Pauletta Hansel's nine poetry collections include *Heartbreak Tree*, a poetic exploration of the intersection of gender and place in Appalachia, and *Palindrome*, winner of the 2017 Weatherford Award for Appalachian poetry. She was 2022 Writer-in-Residence for the Public Library of Cincinnati and Hamilton County and Cincinnati's first Poet Laureate, 2016–18. Her writing has been featured in *Oxford American, Rattle, Appalachian Review, Pine Mountain Sand & Gravel, American Life in Poetry,* and *Poetry Daily*, among others.

Marc Harshman's *Woman in Red Anorak* won the Blue Lynx Prize. His fourteenth children's book, *Fallingwater: The Building of Frank Lloyd Wright's Masterpiece* (with coauthor Anna Smucker) was published by Roaring Brook. He is cowinner of the 2019 Allen Ginsberg Poetry Award, and his poem "Dispatch from the Mountain State" was printed in the 2020 Thanksgiving edition of the *New York Times*. *Dark Hills of Home* came out in 2022 to celebrate Harshman's tenth anniversary as Poet Laureate of West Virginia.

Kelli Hansel Haywood is a writer and yoga instructor from Letcher County, Kentucky. She is the author of *Sacred Catharsis: A Personal Healing Journey amidst the Forced Pause of Pandemic* (Belle History, Spring 2021). Other poetry and prose by Kelli appear in publications such as *Appalachian Reckoning: A Region Responds to* Hillbilly Elegy (WVU Press), *Pine Mountain Sand & Gravel, Still: The Journal,* and *Women Speak* (the Women of Appalachia Project).

Melissa Helton is literary arts director at Hindman Settlement School and has been part of the HSS writing community since 2015. Her work has been published in *Shenandoah, Still: The Journal, Anthology of Appalachian Writers, Appalachian Review, Norwegian Writers Climate Campaign,* and more. Her chapbooks include *Inertia: A Study* and *Hewn*. She previously was tenured associate professor of English at Southeast Kentucky Community and Technical College.

A native of upper East Tennessee, **Jane Hicks** is a poet, teacher, and quilter. Her poetry appears in both journals and numerous anthologies, including *Southern Poetry Anthology: Contemporary Appalachia* and *Southern Poetry Anthology: Tennessee*. Her first book, *Blood and Bone Remember*, was nominated for and won several awards. The University Press of Kentucky published her latest poetry book, *Driving with the Dead*, which won the Appalachian Writers Association Poetry Book of the Year Award for 2014. A third book is coming in 2024.

Contributors

Pamela Hirschler grew up in eastern Kentucky, studied creative writing at Morehead State University, and received an MFA in poetry from Drew University. Her poetry and reviews have previously appeared in *Pine Mountain Sand & Gravel*, *Still: The Journal*, *Tupelo Quarterly*, and other journals. Her first poetry chapbook, *What Lies Beneath*, was published in 2019 by Finishing Line Press.

Scott Honeycutt grew up in Virginia and Tennessee. He spent the summer of 1992 working in tobacco fields around Carrolton, Kentucky, and those days furnished some of his best memories. He has published numerous poems, including two chapbooks, *This Diet of Flesh* (2016) and *Twelve Miles North of the Kentucky River* (2018). When Scott is not teaching, he enjoys hiking the hills of Appalachia and spending time with his daughters.

Silas House is the *New York Times* best-selling author of seven novels. He is the recipient of the Southern Book Prize, the Duggins Prize, the Appalachian Book of the Year, the Storylines Prize from the New York Public Library/NAV Foundation, and many other honors. His work has recently appeared in *Time*, the *Atlantic*, the *Washington Post*, the *Bitter Southerner*, *Ecotone*, and many other publications. He is a native of southeastern Kentucky and now lives in Lexington. His tenure as Kentucky's poet laureate began in 2023.

A native of East Tennessee, **Patricia Hudson** credits her writing career to encouragement she received from Harriette Arnow at the Appalachian Writers' Workshop in 1983. She spent a decade as a contributing editor for *Americana* magazine and wrote for *Southern Living*. Her work to preserve Appalachia's mountains was highlighted in *Something's Rising*, by Silas House and Jason Howard. She holds an MFA from Spalding University. Her first novel, *Traces*, was published in 2022.

Tia Jensen is writer, cancer survivor, and chimera. She has worked with national and regional news producers to broadcast her bone marrow transplant story. Advocating for Be the Match and contributing to the *Red Cross Blog*, she encourages blood and stem cell donations. Tia, a Pushcart and Best of the Net nominee, has been published in *Still: The Journal*, *New Southerner*, and *Ruralite Magazine*.

Shelly Jones, a native of western North Carolina, lives, writes, and teaches in Louisville. She holds degrees from the Bread Loaf School of English and the University of North Carolina. Her work has appeared in the Kentucky Girlhood Project, a statewide contemporary art exhibit promoting Kentucky women in the arts, and she is the winner of a graduate fiction-writing prize at Eastern Kentucky University. Her work explores place, religious trauma, gender, and identity.

Contributors

Leatha Kendrick is the author of five poetry collections, most recently *And Luckier* (Accents, 2022). Her poems, essays, and memoirs have been collected in anthologies, including *Listen Here: Women Writing in Appalachia*; *The Southern Poetry Anthology*, volume 3; and *What Comes Down to Us: Twenty-Five Contemporary Kentucky Poets*. She and George Ella Lyon coedited *Crossing Troublesome: Twenty-Five Years of the Appalachian Writers Workshop*. She mentors and leads workshops at conferences, at universities, and for the Carnegie Center for Literacy and Learning (Lexington, Kentucky). http://leathakendrick.com.

Amelia Kirby has worked for the last two decades in Appalachian social justice and community building. She is from Dungannon, Virginia, and lives in Harlan, Kentucky.

Patsy Kisner is the fifth generation of her family to live along the waters of Sycamore Creek in central West Virginia. Her poems have appeared in *Still: The Journal*, *Pine Mountain Sand & Gravel*, *Appalachian Journal*, and the *Women Speak* anthologies. Her poetry chapbook, *Inside the Horse's Eye*, and her poetry collection, *Last Days of an Old Dog*, were published by Finishing Line Press.

Sonja Livingston is the author of four books of nonfiction, including *Ghostbread*, a memoir of childhood poverty and winner of the AWP Prize for Nonfiction that has been widely adopted for classroom use. Other honors include a New York State Arts Fellowship, an *Iowa Review* Award, and a VanderMey Nonfiction Prize. Sonja splits her time between New York State and Virginia, where she is associate professor of creative writing at Virginia Commonwealth University.

Courtney Lucas is a 2017 graduate of Centre College and a 2022 graduate of the University of Idaho's Creative Writing Master of Fine Arts program. She has published in *Still: The Journal* and the *Washington Post*. She was raised in Pikeville, Kentucky, but currently resides with her partner, Noah, in Johnson City, Tennessee, where she works as an Education Abroad Coordinator at East Tennessee State University.

George Ella Lyon is a freelance writer and teacher rooted in family and a wide community of writers and artists. Raised in the mountains of Kentucky, with Ruth Stone, Virginia Woolf, and Harriette Arnow as word-mothers, she writes poetry, picture books, fiction, nonfiction, essays, and songs. In 2015–16, Lyon served as Kentucky Poet Laureate, and in 2022, she was inducted into the Kentucky Writers Hall of Fame. She has taught at the AWW almost every summer since 1980.

Maurice Manning has published seven books of poetry. His most recent book, *Snakedoctor*, was published in 2023. A former Guggenheim Fellow, Manning teaches at Transylvania University and lives with his family on a small farm in Washington County.

Born and raised in the mountains of North Carolina, **Meredith McCarroll** is writer, editor, and educator. Her work has appeared in *Bitter Southerner*, *Avidly*, *Southern Cultures*, *Still: The Journal*, *Cutleaf*, and elsewhere. McCarroll is the author of *Unwhite: Appalachia, Race, and Film* (University of Georgia Press) and coeditor (along with Anthony Harkins) of *Appalachian Reckoning: A Region Responds to "Hillbilly Elegy"* (West Virginia University Press). She lives in Portland, Maine.

Christopher McCurry is the author of *Open Burning* (Accents Publishing). He is a graduate of the Bread Loaf School of English at Middlebury College and a high school English teacher. In 2015, Christopher cofounded Workhorse, a publishing company and community for working writers. He believes that everyone should write poems and that everyone can. You can find him online at workhorsewriters.com.

Ouita Michel is an eight-time James Beard Foundation Award nominee, including nominations for Outstanding Restaurateur and Best Chef Southeast. Michel and her restaurants are regularly featured in local and national media, such as the *New York Times*, *Southern Living*, *Garden & Gun*, Food Network, and the Cooking Channel. She was a guest judge on Bravo's *Top Chef* series. She lives in Midway, Kentucky, where she oversees her newest venture, hollyhillandco.com.

Jim Minick is the author or editor of eight books, including *Without Warning: The Tornado of Udall, Kansas* (nonfiction), *The Intimacy of Spoons* (poetry), *Fire Is Your Water* (novel), and *The Blueberry Years: A Memoir of Farm and Family*. His work has appeared in many publications, including the *New York Times*, *Poets & Writers*, *Oxford American*, *Orion*, *Shenandoah*, *Appalachian Journal*, *Wind*, and the *Sun*. He serves as coeditor of *Pine Mountain Sand & Gravel*.

Britton Patrick Morgan is a Kentucky singer, songwriter, and producer. He has collaborated or produced works for numerous Kentucky and Appalachian artists, including Tiffany Williams, Darrell Scott, Nolan Taylor, Malcolm Holcombe, and Bill Alexander. Morgan also dedicates time to the preservation of Kentucky's natural and folk treasures, most recently spearheading the successful effort to acquire and restore Harlan and Anna Hubbard's home, Payne Hollow, on the banks of the Ohio River in Milton, Kentucky.

Lisa J. Parker is a native Virginian, a poet, a musician, and a photographer. Her first book, *This Gone Place*, won the 2010 ASA Weatherford Award; her second book, *The Parting Glass*, won the 2021 Arthur Smith Poetry Prize; and her work

Contributors

is widely published in literary journals and anthologies. Her photography has been on exhibit in New York City and published in several arts journals and anthologies. Some of her work may be found at www.wheatpark.com.

Tina Parker is the author of the poetry collections *Lock Her Up, Mother May I,* and *Another Offering*. Tina grew up in Bristol, Virginia. She is a longtime Kentucky resident and advocates for poetry across the state: she is the former President of the Kentucky State Poetry Society and a juried member of the Women of Appalachia Project.

Julie Rae Powers received their MFA in Photography from Ohio State University and their BFA in Photography from James Madison University. Their photographic and written work has focused on family history, coal, Appalachia, queerness, and the butch body.

Melva Sue Priddy is the author of *The Tillable Land* (2021), a memoir-in-verse about her childhood and growing-up years spent laboring on her family's dairy and tobacco farm. Her poems witness survivance, growth, and the healing power of feet in the soil. A native of Kentucky, she started attending HSS Appalachian Writers' Workshop while attending Berea College with two children.

Erin Miller Reid is a physician in Kingsport, Tennessee. Her short story "Uncaged" won the 2022 Doris Betts Fiction Prize and was published in *North Carolina Literary Review*'s 2023 summer volume. Another short story, "The Offering," won first place in *Still: The Journal*'s 2018 fiction contest. She has also been published in *Appalachian Review, The Women of Appalachia Project Anthology, Pine Mountain Sand & Gravel,* and *100 Days in Appalachia*. She has completed her first novel.

Amy Le Ann Richardson was born and raised in Morehead, Kentucky, and holds an MFA from Spalding University (2009). Amy is a farmer, writer, visual artist, and teacher and has received grants and fellowships from the Kentucky Foundation for Women. She is the author of *Who You Grow Into*, and her work has been featured in *Pine Mountain Sand & Gravel*, the *Yearling*, and *Kentucky Monthly*. She lives and works on her farm in Carter County.

Mandi Fugate Sheffel was born and raised in Red Fox, Kentucky. A graduate of Eastern Kentucky University, she found her passion for writing and storytelling at the Appalachian Writers' Workshop. Her essays and opinion pieces can be found in *Still: The Journal, Lexington Herald-Leader,* and the *Louisville Courier Journal*. Her forthcoming personal essay collection will be published by the University Press of Kentucky. She owns and operates Read Spotted Newt, an independent bookstore in the coalfields of eastern Kentucky.

Contributors

Carter Sickels is the author of the novels *The Prettiest Star* and *The Evening Hour*. His writing appears in various outlets, including the *Atlantic*, *Oxford American*, *Poets & Writers*, *BuzzFeed*, *Guernica*, and *Joyland*. He has received fellowships from the Bread Loaf Writers' Conference, the Sewanee Writers' Conference, MacDowell, and the Virginia Center for the Creative Arts. You can find more information at www.cartersickels.com.

Savannah Sipple is the author of *WWJD & Other Poems* (Sibling Rivalry Press, 2019), which was included on the American Library Association's Over the Rainbow Recommended LGBTQ Reading List. She is a writer from eastern Kentucky, and her writing has been published in *Salon*, *Go Magazine*, and other places. A professor, editor, and writing mentor, Savannah resides in Lexington with her wife.

Amanda Jo Slone lives and writes in Draffin, Kentucky. She serves as assistant provost and professor of English at the University of Pikeville. Slone earned her MFA in Creative Writing from West Virginia Wesleyan College and PhD in Educational Leadership at Northwest Nazarene University. Her work has appeared in *Still: The Journal*, *Appalachian Review*, *Louisville Review*, and other journals and anthologies.

Lee Smith is the author of fifteen novels, including *Fair and Tender Ladies*, *Oral History*, *Saving Grace*, and *Guests on Earth*, as well as four collections of short stories. Her novel *The Last Girls* was a *New York Times* bestseller as well as cowinner of the Southern Books Critics Circle Award. A retired professor of English at North Carolina State University, she has received an Academy Award in Fiction from the American Academy of Arts and Letters, the North Carolina Award for Literature, and the Weatherford Award for Appalachian Literature.

Jamey Temple is a writer and professor who teaches English at University of the Cumberlands in eastern Kentucky. Her poetry and prose have been included in several publications, such as *River Teeth*, *Rattle*, *Appalachian Review*, *Bending Genres*, and *Still: The Journal*. She was named finalist in *Fourth Genre*'s 2022 Multimedia Essay Prize and finalist for *Newfound Journal*'s Prose Prize in 2016. You can read more of her published work through her website, jameytemple.com.

Lyrae Van Clief-Stefanon is the author of *] Open Interval [*, a finalist for the National Book Award and the *LA Times* Book Prize, and *Black Swan*, winner of the Cave Canem Poetry Prize. She has been awarded fellowships from Cave Canem, the Lannan Foundation, and Civitella Ranieri. She is associate professor in the Department of Literatures in English at Cornell University.

Doug Van Gundy directs the Low-Residency MFA program in Creative Writing at West Virginia Wesleyan College. His poems, essays, and reviews have appeared

Contributors

in many journals, including *Poetry, Poets & Writers*, and the *Oxford American*. He is coeditor of *Eyes Glowing at the Edge of the Woods: Contemporary Writing from West Virginia* and the author of a book of poems, *A Life Above Water*. He lives in West Virginia.

Jayne Moore Waldrop is a Kentucky writer and attorney. Her books include *Drowned Town* (University Press of Kentucky), a 2022 Great Group Reads selection by the Women's National Book Association; *A Journey in Color: The Art of Ellis Wilson*; *Pandemic Lent: A Season in Poems*; and *Retracing My Steps*, a New Women's Voices Chapbook Series finalist. Her work has appeared in *Appalachian Review*, *Still: The Journal*, *Anthology of Appalachian Writers*, and *Women Speak* anthologies.

The first African American writer to be named Kentucky Poet Laureate, a multidisciplinary artist, a writer, and an educator, **Frank X Walker** has published eleven collections of poetry and a new children's book, *A Is for Affrilachia*. He is professor of English and African American and Africana Studies and Director of the MFA in Creative Writing program at the University of Kentucky. He is a Cave Canem Fellow, and his honors also include a Lannan Literary Fellowship for Poetry.

Randi Ward is a poet, translator, lyricist, and photographer from West Virginia. She earned her MA in Cultural Studies from the University of the Faroe Islands and has twice won the American-Scandinavian Foundation's Nadia Christensen Prize. Ward's work has been featured on Folk Radio UK, NPR, and PBS NewsHour. Cornell University Library established the Randi Ward Collection in its Division of Rare and Manuscript Collections in 2015. For more information, visit randiward.com/about/.

A native of southeastern Kentucky, **Julia Watts** is the author of the young adult novels *Needlework* and *Quiver*, as well as a lot of other novels for adults and young adults. Her novel *Finding H.F.* won a Lambda Literary Award. She lives in Knoxville and is working on a PhD in Children's and Young Adult Literature at the University of Tennessee. Her novel, *Finding H.F.*, won a Lambda Literary Award.

Annie Woodford is from a mill town in the Virginia Piedmont. She is the author of *Bootleg* (Groundhog Poetry Press, 2019). Her second book, *Where You Come from Is Gone* (2022), is the winner of Mercer University's 2020 Adrienne Bond Prize and the 2022 Weatherford Award for Appalachian poetry. Her work has appeared or is forthcoming in *Cutleaf*, *Still: The Journal*, *Appalachian Journal*, and *Appalachian Heritage*, among others.

Contributors

Marianne Worthington is author of *The Girl Singer* (University Press of Kentucky, 2021), winner of the 2022 Weatherford Award for poetry. Her work has appeared in *Oxford American*, *CALYX*, *Chapter 16*, and *Cheap Pop*, among other places. She cofounded and is poetry editor of *Still: The Journal*, an online literary magazine publishing writers, artists, and musicians with ties to Appalachia since 2009. She grew up in Knoxville, Tennessee, and lives and teaches in southeastern Kentucky.

Made in United States
Cleveland, OH
19 May 2025